Haldane's *Daedalus* Revisited

Biochemistry Department, University of Cambridge — The Staff, 1930

Back row

Ruby Leader, Mrs H. Smith, Eric Benton, Douglas Dewey, Alfred Cowell, Fred Johnson, Fred Jolley, Basil Slater, Alfred Colwell, Rhoda Vincent

Middle row

Lily Gordon, Olive Impey, Henry Mowl, Henry W. Hall, 'Charles' Jolley, J.R.C. Powney, George Leader, E.J. Morgan, G.W. Naylor, Stanley Williamson, Joan Coard

Front row

Barbara Callow, Dorothy Needham, Joseph Needham, Malcolm Dixon, J.B.S. Haldane, *The Professor* (Sir Frederick Gowland Hopkins), S.W. Cole, T.S. Hele, Muriel Wheldale Onslow, Marjory Stephenson, Robin Hill

HALDANE'S *DAEDALUS* REVISITED

Edited with an introduction by

KRISHNA R. DRONAMRAJU

Foreword by

JOSHUA LEDERBERG

Oxford New York Tokyo
OXFORD UNIVERSITY PRESS
1995

Oxford University Press, Walton Street, Oxford OX2 6DP
Oxford New York
Athens Auckland Bangkok Bombay
Calcutta Cape Town Dar es Salaam Delhi
Florence Hong Kong Istanbul Karachi
Kuala Lumpur Madras Madrid Melbourne
Mexico City Nairobi Paris Singapore
Taipei Tokyo Toronto
and associated companies in
Berlin Ibadan

Oxford is a trade mark of Oxford University Press

Published in the United States
by Oxford University Press Inc., New York

A catalogue record for this book is available from the British Library

Library of Congress Cataloging in Publication Data
Haldane's Daedalus revisited / edited with an introduction by Krishna R. Dronamraju; foreword by Joshua Lederberg.
Includes reprint of a great classic, 'Daedalus or Science and the Future'.
Includes bibliographical references and index.
1. Science. 2. Biology. I. Dronamraju, Krishna R. II. Haldane, J. B. S. (John Burdon Sanderson), 1892–1964. Daedalus. III. Title: Daedalus revisited.
[DNLM: 1. Haldane, J. B. S. (John Burdon Sanderson), 1892–1964. Daedalus.]
Q171.H1566 1995 303.48'3–dc20 94-34310
ISBN 0 19 854846 X

Typeset by Downdell, Oxford
Printed in Great Britain by
Biddles Ltd, Guildford & King's Lynn.

Dedication

This book is dedicated to the memory of J.B.S. Haldane whose writings have deeply influenced and stimulated generations of scientists and writers. In his introduction to *Daedalus* Haldane emphasizes that it is the 'whole business of a university teacher to induce people to think'. This dedication expresses in some small way the gratitude of myself and others for the intellectual stimulation which we have received from Haldane's writings.

Foreword

Joshua Lederberg

J.B.S. Haldane's *Daedalus* 1923—70 years before and after

70 years after the publication of *Daedalus* in 1923, most of Haldane's themes are intelligible, even familiar, however we may criticize them in detail. 1923 is separated by a much deeper gulf from 1853. Despite the enormous acceleration of science and technology in this septuadecennium, there were far more deepseated revolutions of thought in the previous one: evolution, infection, the gene, molecular organics, the radioactive atom, electromagnetic radiation, relativity, quantum theory.

From the vantage point of 1923, Haldane could but dimly anticipate the further directions of science, but could more self-assuredly extrapolate the technological applications of some of these breakthroughs. Inevitably, he missed many boats: nuclear energy, the electronic computer, space travel. He'd had radioactivity in mind, but could not foresee the neutron, nuclear fission, or fusion. Ectogenetic procreation—prefiguring Aldous Huxley's *Brave new world*—is virtually achieved; its near equivalents of *in vitro* fertilization, and embryo culture and transplantation are in today's headlines. And the separation of sexual gratification from reproduction is here to stay.

Haldane also inspired Huxley's vision of 'soma', the innovation of mind-altering chemistry, for which LSD and Prozac are the crude beginnings. By meeting universal wants, a 'safe' soma may be the most devastating of the biological technologies on the horizon—matched only by the promise of indefinite life-extension, at infinite cost.

Can one do any better in anticipation from 1994? But first, what merit is there to such an exercise? With politics, via the multimedia, so inextricably fused with entertainment, I despair that any more rational planning could ensue. In fact, so few people can admit the possibility of any change of moral or social perspective as to invite a kind of imperialism of the present over the future, a closing of options that our children should have available to them. And they would argue in turn that they receive a legacy of technology, about whose merits they had no voice in deciding. So our prior ethical task is to outline what is owing in intergenerational responsibility—one beyond the scope of this essay.

Today's nightmares embrace population, pollution, plagues, and proliferation of weapons. For the first three, technology has much to offer in mitigation, and rational explication does play some role in engendering the political will to pay for the remedies. For the fourth, nothing would be better than an enforceable moratorium. Leo Szilard once said, 'The optimist is one who believes the future is uncertain', and to that degree there remains a shred of hope for the needed world order. Scientists are the most cosmopolitan agents in a contemporary culture that becomes ever more particularistic, including in the exploitation of the fruits of science and technology.

Technological futurism is so widely practised today; I will not dwell on a landscape oft painted by others. Space travel, computers, multimedia, and the global village are all familiar themes, and nevertheless quite realistic. In fact other physical technologies may be accelerated with the truncation of effort in mega-high-energy particles, which have the remotest bearing on everyday existence. Within biology, working out the DNA paradigm (Lederberg 1993) is enough to occupy several generations of researchers, and to offer zones of application in technology at least as immense as its proponents advertise—not necessarily ensuring profits for the specific ventures currently touted. The last bastions of biological mystery, embryonic development and the conscious brain, have a century's trove of secrets to be plumbed. How we define individual personality becomes the next ethical challenge that must confront materialistic biological fact. The DNA paradigm is so deep, so pervasive, that it has become difficult to speculate on what could ever lie beyond it—at the very horizon, just how life could have evolved on earth, or in cosmological differentiation before that. And therefrom, the actual synthesis of alternative life forms, based on chemistries other than DNA, and imaginably already extant elsewhere in the universe, and accessible only by telecommunication. Biology is already so fact laden that it is in danger of being bogged down awaiting advances in logic and linguistics to ease the integration of the particulars. So we welcome better esoteric communication as well. This very text stands a better chance of some useful influence, or inviting proper criticism, when it becomes electronically available 'online'.

Haldane's writing sits athwart sceptical utopias—he bridges Samuel Butler and Aldous Huxley—and he sheds only a trace of optimism that human political arrangements will successfully master technological power for broader advantage. '. . . the tendency of applied science is to magnify injustices until they become too intolerable to be borne. . . .'

Public grievances about science and technology have become even more articulate since Haldane's writing—though we recall that Shelley's *Frankenstein—the modern Prometheus* dates to 1818, and is derived from classical and Old Testament allusions. In an increasingly technological civilization, one could characterize public perceptions as being more passionately ambivalent: full of deeper fear, dependence, expectation, resentment, and incomprehension (Lederberg 1972). In 1923, Haldane still railed against religious superstition and bigotry—this is perhaps coming back full circle. But for the most part, science has played its part in the attrition of the old faiths. In 1923 science could be viewed as a counter-religion; but this movement has failed utterly. Up through the nineteenth century, science could offer rational explanation of many features of everyday life: the movements of the planets, fire, electricity, healing drugs, biological evolution and diversity, in terms that enriched lay understanding and a sense of personal control of the environment. This is no longer true: the environment has become infinitely more complex, and the cutting edge of scientific discovery needs tomes of background just to understand the conceptual issue. And that technological environment is changing so rapidly, and posing so many burdens of decision, that everyman does become bewildered, resentful, and mistrustful of established authority. Above all, science is bereft of deontology: it cannot tell why one should be interested in science or anything else. It is no replacement for the anchor of religious faith; at its best it can help explore the consequences of our beliefs and actions, and complicate our ethical judgments accordingly.

REFERENCES

Lederberg, J. (1972). The freedom and the control of science—notes from the ivory tower. *Southern California Law Review*, **45**, 596–614.

Lederberg, J. (1993). What the double helix (1953) has meant for basic biomedical science: a personal commentary. *Journal of the American Medical Association*, **269**(15), 1981–5.

ACKNOWLEDGEMENTS

I am grateful to all the contributors for their prompt cooperation. I thank Mark Weatherall and Annette Faux for the staff photograph of the Department of Biochemistry, University of Cambridge, taken in 1930. For valuable advice and helpful suggestions, I express my appreciation to Joshua Lederberg, M.F. Perutz, Arthur C. Clarke, N. Avrion Mitchison, Victor A. McKusick, William J. Schull, Cedric A.B. Smith, C.R. Rao, and M.S. Swaminathan.

Permission to reproduce *Daedalus or Science and the Future* by J.B.S. Haldane has kindly been granted by Routledge & Kegan Paul.

ACKNOWLEDGEMENTS

[illegible]

[illegible]

Contents

CONTRIBUTORS

Elof Axel Carlson

Department of Biochemistry and Cell Biology, University at Stony Brook, Stony Brook, NY, USA

Krishna R. Dronamraju

Foundation for Genetic Research, P.O. Box 27701-497, Houston, TX 77227, USA

Freeman Dyson

Institute for Advanced Study, Princeton, NJ, USA

Yaron Ezrahi

Department of Political Science, The Hebrew University of Jerusalem, Jerusalem, Israel

Joshua Lederberg

The Rockefeller University, New York, USA

Ernst Mayr

Museum of Comparative Zoology, Harvard University, Cambridge, MA 02138, USA

N.A. Mitchison

Deutsches Rheuma Forschungs Zentrum, Nordufer 20, D-13353 Berlin, Germany

M.F. Perutz

MRC Laboratory of Molecular Biology, Hills Road, Cambridge, CB2 2QH, UK

D.J. Weatherall

Institute of Molecular Medicine, University of Oxford, Oxford, UK

CHAPTER ONE

INTRODUCTION

KRISHNA R. DRONAMRAJU

A scientist must be at once a dog with his nose to the ground, and a god using the purest form of reason

J.B.S. Haldane

J.B.S. Haldane was 31 years old when he read a seminal paper on the future of science and its applications to the Heretics at Cambridge, on 4 February 1923. It was subsequently published as a book under the title *Daedalus, or science and the future* (in 1923) by Kegan Paul in London, and a year later in New York by E.P. Dutton (Haldane 1924*a*). After its first printing in New York in April 1924, it was reprinted five more times that year, three times in 1925, and there was a tenth printing in 1926. During the first year, about 15 000 copies were sold.

The following is a brief history of events which led up to the publication of *Daedalus*. While he was an Oxford undergraduate in 1912, Haldane wrote an essay dealing with the future applications of science but did not pursue it further until he was invited nine years later by the New College (Oxford) Essay Society. He refurbished it in the light of scientific advances made during the intervening years and his own wartime experiences which included the new and deadly chemical warfare. Shortly afterwards, in 1923, when he was invited to address the Heretics at Cambridge, Haldane once again brought it up to date. His audience included C.K. Ogden, the creator of Basic English and a former editor of *The Cambridge Magazine*. As a 'scout' for the publisher, Kegan Paul, Ogden encouraged Haldane to write up his lecture in a publishable form to be included in the new series *Today and Tomorrow*, which his firm was about to launch. Throughout his life, Haldane repeatedly wrote on the subject of eugenic selection, genetic engineering and its consequences, and related topics; but the passionate and bold imagination which he displayed so clearly in *Daedalus* was never to be seen again. In his introduction to *Daedalus*, Haldane wrote that 'it is the whole business of a university teacher to induce people to think.' He achieved that goal admirably.

CONTROVERSY

Almost from the instant it appeared, *Daedalus* caused controversy. It dealt with revolutionary biological intervention at a time when even the mention of 'birth control' in public media caused an uproar. As late as 1926, Julian Huxley was rebuked by the Head of British Broadcasting Corporation (BBC), Lord Reith, for mentioning 'birth control' on the BBC radio (Dronamraju 1993). Margaret Sanger opened the first birth control clinic in North America in Brooklyn in 1916, and Marie Stopes established a similar one in London shortly afterwards. These were centres of much public controversy and conflict. Haldane's *Daedalus* appeared in the midst of these sweeping social changes which were being spearheaded by the suffragette movement in England and America. Haldane predicted (or advocated) not merely 'birth control', which was repugnant enough to most people at that time, but a direct and active program of intervention to manipulate the human genome—eugenic selection, *in vitro* fertilization, and routine production of individuals with exceptional qualities in music, sports, and virtue. Furthermore, he prophesied the widespread use of psychotropic drugs and numerous other biological and therapeutic interventions to modify the psychological and behavioural, as well as the biochemical, condition of the human body and mind. *Daedalus* quickly became the staple subject for gossip in the college halls and the tea shops around Oxford. The notoriety stirred up by Haldane's biological predictions caused some anguish to his parents. In his *Memories*, Julian Huxley (1970) quoted the following letter from Mrs. Louisa Kathleen Haldane (J.B.S. Haldane's mother):

Cherwell, Oxford

Dear Julian

I find the SP (she always called her husband the Senior Partner) is frightfully upset about *Daedalus*. Will you abstain altogether from poking fun at him on account of it? and if you can do so, keep people off the subject altogether when he is about?

I knew h'd object, but had no idea till to-day how really unhappy he is—odd people these Liberals and no accounting for them!

But an' you love me, keep people off him, or he'll hate you all! (which is not only sad for him but extremely inconvenient for me.)

Yours aff:
LKH.

Haldane, of course, immensely enjoyed the sensation it caused and took great pleasure in expounding his views loudly at every opportunity, quickly clearing tea shops of the more susceptible churchgoers. His great skill at scientific popularization quickly became evident, establishing him as one of the most eminent scientist-writers for the rest of his life. *Daedalus* was soon followed by a number of popular essays which were later published in a book entitled *Possible worlds and other essays* (1927). This collection contains one of Haldane's most famous essays, 'On being the right size', which was described by Arthur C. Clarke (1968, p. 243) as a 'perfect example of his lucidity and the breadth of his interests'. It was also in 1927 that a textbook, *Animal biology*, appeared with Haldane and Julian Huxley as co-authors. Haldane continued to popularize science throughout his life, contributing numerous articles to the popular press in several countries. He continued popular writing in India where he lived during his last years, 1957–64 (Dronamraju 1985, 1987).

In response to Haldane's *Daedalus* (1923), Bertrand Russell's *Icarus, or the future of science* (1924) appeared in the following year. Russell disagreed with the benevolent view of science presented by Haldane, arguing that science is not a substitute for virtue and reason. According to Russell, technological applications make the world less safe, not more secure.

It was typical of Haldane that he attracted (or sought after) controversy for much of his life. A few years after its publication, he was embroiled in a major controversy about *Daedalus* with the ageing F.E. Smith, first Lord Birkenhead and former Lord Chancellor and Secretary of State for India. In 1930, Birkenhead's book *The world in 2030* appeared, and might have passed almost unnoticed except for the fact that it happened to reach Haldane, who reviewed the book in the *Weekend Review* under the heading 'Lord Birkenhead improves his mind'. Haldane commented, 'Certain of the phrases seemed unduly familiar. Where had I seen them before? Finally I solved the mystery. They were my own.' Haldane stated that he had noted a total of forty similarities between Birkenhead's book and *Daedalus*. Haldane concluded 'I will not say . . . plagiarism, but . . . a certain lack of originality . . . because it carries with it corollaries which I find unthinkable.' Haldane even suggested a tongue-in-cheek explanation, that both he and Birkenhead must have seen the same original—including one written 48 years hence!

Now a wiser man would have let the matter drop at that stage because Haldane contemplated no legal recourse, but not Lord

Birkenhead. The latter could have pleaded an inadvertent oversight as a possible explanation. Instead, Birkenhead responded by attacking Haldane in the columns of one of the most widely circulated newspapers, the *Daily Express*, which naturally led to the sudden popularity of *Daedalus* among the general public, who would otherwise have remained blissfully ignorant of its existence. Birkenhead claimed that Haldane lacked historical knowledge and that he (Birkenhead) was following in other men's footsteps. Haldane retorted (in the *Weekend Review*) that his knowledge of history had earned him a first in literae humaniores at Oxford. He noted further that he had no objection to Birkenhead following in other men's footsteps but that 'I object to . . . stealing my boots to do so, and I am amused when they do not know how to put the boots on.'

That was the end of the controversy, one of many in J.B.S. Haldane's life. He enjoyed them all (Dronamraju 1985).

SCIENTIFIC BACKGROUND

John Burdon Sanderson Haldane was born in Oxford, England, on 5 November 1892, and died in Bhubaneswar, India, on 1 December 1964. J.B.S. (or Jack) grew up in that intellectually privileged class of liberal intellectuals which included George Bernard Shaw, H.G. Wells, Bertrand Russell, Julian Huxley and his younger brother Aldous, D.H. Lawrence, and many others who spearheaded the twentieth century's intellectual emancipation. Jack's father was John Scott Haldane, eminent Oxford physiologist who was well known for his physiological experiments in respiration (Clark 1968; Dronamraju 1968). The younger Haldane was devoted to his father, who taught him the essentials of scientific research (Mitchison 1968). Naturally, it was thought at first that Jack would become a physiologist in his father's footsteps. Indeed, J.B.S. Haldane's first scientific paper was in respiratory physiology in collaboration with his father and C.G. Douglas (Douglas *et al.* 1912). However, Haldane's formal education at Oxford was in the humanities. He graduated with honours in 1914, but earned no academic degree in any branch of science. Simultaneously, he initiated breeding experiments in collaboration with his sister Naomi (later Lady Mitchison) and A.D. Sprunt to study what was called 'reduplication' (later 'linkage') in mice (Haldane *et al.* 1915). This work was interrupted by world war I during which he served in the Black Watch battalion with exceptional bravery. Returning to Oxford after world war I, Haldane joined New College as a Fellow in physiology and, in 1919, published an important paper on the estimation of map distances between loci based on recombination frequency. This was the first attempt to develop a mapping function. In the same paper, Haldane proposed the term 'centimorgan' (cM) as a measure of gene mapping.

Haldane was a pioneer in biochemical genetics (Haldane 1920, 1937). With reference to the biochemical nature of the gene and gene action, Haldane (1920) wrote: 'The chemist may regard them (genes) as large nucleoprotein molecules, but the biologist will perhaps remind him that they exhibit one of the most fundamental characteristics of a living organism: they reproduce themselves without any perceptible change in various different environments . . . in some cases we have very strong evidence that they produce definite quantities of enzymes, and that the members of a series of multiple allelomorphs produce the same enzyme in different quantities.'

Continuing his genetical studies (under the influence of William Bateson), Haldane, in 1922, published an interesting generalization, later called 'Haldane's rule', which stated that 'When in the F_1 offspring of a cross between two animal species or races one sex is absent, rare, or sterile, that sex is always the heterozygous sex.' Although 'Haldane's rule' is by far the most quoted achievement of Haldane, which is connected with his name in the minds of most students of biology, it was not his best or most important scientific contribution. His greatest contribution may well be his mathematical theory of natural selection, which was largely developed during the years 1924–34 (see Haldane 1924*b*). Haldane's early work in this field was summarized in his book, *The causes of evolution* (1932*a*). Along with the works of R.A. Fisher and S. Wright, this work places Haldane as a founder of population genetics and as an eminent leader of twentieth century biology, especially theoretical biology. What makes Haldane's early papers even more significant than those of his great contemporaries is the great breadth of his knowledge of not only the mathematics of genetics and evolution, but also of biochemistry and physiology, and the bold extrapolation and synthesis of ideas which they contain—ranging from biology to chemistry, physics, and astronomy, and also theology and ethics as well as science fiction (see Dronamraju 1989, 1990, 1992; Crow 1984). This was especially evident in *Daedalus*.

Daedalus was preceded by only fifteen papers, mostly in human physiology—especially the physiological effects of raising or lowering the blood pH. What is interesting is that *Daedalus* represents a landmark in Haldane's writings. It was the kind of bold speculative writing which characterized much of his writing for the rest of his life. But why *Daedalus*, and why did it appear at that particular time in Haldane's life? Until then, his publications were typical of what one would expect of any other university teacher. The physiological papers appear to be a continuation of the kind of research which his father followed for many years. But the two genetical studies (published in 1914 and 1919) were clearly much more original and important: first, a discovery of linkage in the vertebrates; second, the development of the first mapping function. So it is obvious that his heart (and talent) lay in genetics, not in physiology.

Although continuing to tread the path laid down by his father in physiology at first, Haldane appears to have been keenly interested in making a mark of his own, in a different branch of science, and the young science of genetics provided that opportunity. This was perhaps his way of establishing his own identity as a scientist, separate from his father's, which was to be expected in a young scientist, especially one with a famous scientist-father. In 1901, when Haldane was only eight years old, his father took him to a Royal Society conversazione where Haldane heard A.D. Darbishire lecture on the recently rediscovered work of the Austrian monk, Gregor Mendel (Haldane 1960, personal communication). That particular encounter with the beginnings of genetics struck an intellectual chord in young Haldane's mind. It provided him an alternative intellectual interest (and excitement) to the physiological experiments of his father which he already knew. Several years later, writing about the scientific method, Haldane (1963*a*) stated: 'a piece of research directed by a good scientist should leave one with high standards of accuracy and integrity which one can transfer to other fields of science. . . . I think that for most of us an occasional change is desirable because we are apt to think that the topics which, very rightly, excited us in our twenties, are still the most important.'

SOCIAL CLIMATE

It is interesting to speculate the possible causes which led Haldane to write *Daedalus*. Until then his published scientific work was not particularly concerned with eugenics or scientific predictions. This was not the case with some of his contemporaries such as R.A. Fisher, Julian Huxley, and H.J. Muller, who retained a lifelong interest in eugenics.

Daedalus was Haldane's first publication of his eugenical views. It is of interest that he chose to do so as soon as he moved to Cambridge, away from New College, Oxford, where he was a Fellow in physiology, still deeply under his father's influence. While his early encounters with Mendelian genetics and later appreciation of the fundamental nature of the biochemical basis of gene action may have led to eugenical speculations, the social and political milieux of those years may also have had a decided impact on Haldane.

Lady Ottoline Morrell

A possible explanation may lie in the events and the cast of characters which surrounded Haldane during the years preceding and following world war I. Haldane and his sister Naomi (Lady Mitchison) were inevitably drawn into wider intellectual circles, starting with their close friends the Huxleys—Julian and Aldous. Naomi wrote and produced plays. In her contribution to my book *Haldane and modern biology* (1968), she wrote: 'I dragged them all into acting, for at that time I was forever writing plays, first *Saunes Bairos* about an imaginary country in the Andes where eugenics was highly organized by the priesthood.' Now let us compare this sentence with the following sentence from *Daedalus* (p. 35): 'The eugenic official, a compound, it would appear, of the policeman, the priest and the procurer, is to hale us off at suitable intervals to the local temple of Venus Genetrix with a partner chosen, one gathers, by something of the nature of a glorified medical board. To this prophecy I should reply that it proceeds from a type of mind as lacking in originality as in knowledge of human nature. . . . It is moreover likely, as we shall see, that the ends proposed by the eugenist will be attained in a very different manner.' It appears then that the eugenical ideas which Haldane expressed in *Daedalus* were very much a part of their social and intellectual culture during the preceding several years. In the same essay, Naomi had recorded that her brother was interested in genetics while still at Eton (1905–11).

In their social circles, intellectual boundaries were blurred—pursuit of science, literature, and acting as well as numerous other activities, brought them all together. Naomi was close to the Guilguds as well—both Lewis and John. Through Aldous and Julian they were drawn closer to the intellectual centres of Bloomsbury, especially the left-leaning group which found a generous hostess in Lady Ottoline Morrell—married to politician Philip Morrell, but well known to have many lovers, among them Bertrand Russell and Julian Huxley. She was a superb hostess (at her country home Garsington Manor) to numerous intellectuals including Virginia Woolf, T.S. Eliot, Desmond MacCarthy, Lytton Strachey, Duncan Grant, John Maynard Keynes, Vanessa and Clive Bell, Margot Asquith, Katherine Mansfield, W.B. Yeats, David Cecil, Eddy Sackville-West, Maurice Bowra, Wyndham Lewis, Juliette Huxley (who looked after Ottoline's daughter, Julian), and Augustus John, and of course Bertrand Russell and Julian and Aldous Huxley.

Although Haldane remained on the periphery of that group, the literary and artistic mores of *Daedalus* were no doubt influenced by the intellectual activities and climate of that period. Social emancipation, especially the suffragette movement and sexual liberation, were spearheaded by that group. This change was, in part, due to a rejection of the hypocrisy and guilt associated with sex that was characteristic of the preceding era. Writing of his own experience, Julian Huxley (1970) commented: 'this battle between sexual attraction and a puritanical sense of guilt. . . . The whole climate of the Edwardian age, with its hypocritical suppression of everything "nasty", fostered this conflict between instinct and reason.'

Close friends of Ottoline included D.H. Lawrence and his wife Frieda, both also close friends of Aldous Huxley. Controversy was very much a hallmark of that group. Seymour (1992) has recently presented a comprehensive account of their activities in a fine biography of Ottoline Morrell. Lawrence's book *The rainbow* was published in 1915. One of the characters ridicules a soldier's uniform while the author glorifies lesbian behaviour. And it was dedicated to Lawrence's German sister-in-law. These facts were not congenial to the wartime climate, and the book was widely attacked in the press as a menace to the war effort against Germany. The Director of Public Prosecutions ordered all remaining copies of the book destroyed. The presiding magistrate, who lost his own son on the war front, had little sympathy for Lawrence. The publisher of *The rainbow*, Algernon Methuen, was fined ten guineas and was ordered to withdraw the novel. Haldane's sister, the writer Naomi Mitchison, stated that there was a widespread feeling of rebellion among the young who were mostly supportive of Lawrence and the right of freedom of expression. She was among those present at the trial of Lawrence and his publisher (Mitchison 1992, personal communication).

Some years later, a similar fate awaited Lawrence's *Lady Chatterley's lover*, which was based on his affair with Ottoline. Another of Ottoline's lovers, Bertrand Russell, was a pacifist who went to prison for refusing to serve in the war. On the scientific side, Julian Huxley too stirred up notoriety and controversy when he discovered that the metamorphosis of the axolotls, an amphibian which retains for all its life the gills and the broad swimming fin of its tadpole state, could be speeded up by feeding them on thyroid gland. Huxley's brief note, which appeared in *Nature* in 1920, immediately became the subject for headlines in the popular press. One headline read: 'Young British scientist finds elixir of life'. Some of his friends, including J.B.S. Haldane, warned him of the dangers of publicity. Ironically, it was Haldane's turn to cause notoriety when his *Daedalus* appeared in 1923.

So, in retrospect, *Daedalus* appears to be another landmark in a series of sensational events and publications which emanated from the rebellious left-wing intellectual elite during the pre- and post-world war I years. From the establishment's point of view, these were no doubt outrageously irritating incidents. Clearly there was a connection between the wartime experiences and several publications of that period. This was obvious in the case of *Daedalus*, which opens with a description of the battle scene from world war I.

Haldane's *Daedalus* served several purposes:

(1) viewed against the rebellious climate among the younger generation of that time, it was perhaps Haldane's desire to shock the establishment;
(2) it was a warning against possible misuse or excessive zeal in applying science to solve social problems;
(3) it was an attempt to encourage certain lines of biological research;
(4) it was a prediction of what future scientific discoveries will inevitably bring forth;
(5) it was a comparative discussion of progress in the biological and physical sciences;
(6) it was a discussion of scientific applications in war and peace;

(7) it was an evaluation of the impact of science on our ethical outlook;
(8) it was a combination of all the above.

With respect to H.G. Wells, Haldane stated that the scientific predictions of Wells inspired him to launch his own writing career. His admiration for Wells can be seen in *Daedalus*. Haldane wrote that the very mention of the future suggests the name of Wells. Citing Wells' prophecy (in a book called *Anticipations*, 1902), that by 1950 there would be heavier than air flying machines capable of practical use in war, Haldane promised that he would not make any prophecy that would be rasher than Wells' prediction of flying machines.

While Wells may have inspired Haldane's writing career, scientific knowledge and inspiration were provided also by several other individuals, the most important mentor being his father John Scott Haldane. While his formal education was in the humanities, Haldane enjoyed attending the finals honours course in zoology taught by E.S. Goodrich at Oxford during 1911–12. He took the course almost as a relaxation from his long mathematical labours. Goodrich had an unusual career. He first received training in art at the Slade School in London, but later studied zoology under Ray Lankester. More than fifty years later, Haldane still vividly remembered Goodrich's excellent colour drawings and how they inspired his biological interests (Haldane 1960, personal communication).

SCIENTIFIC PREDICTIONS

In *Daedalus*, Haldane covered far more than biology and eugenics. He considered that coal and oil fields would be exhausted in a few centuries and water power not be a suitable substitute. But he predicted that (p. 30) 'four hundred years hence the power question in England may be solved somewhat as follows: The country will be covered with rows of metallic windmills working electric motors which in their turn supply current at a very high voltage to great electric mains.' (The subject of exploiting wind power to meet energy needs received special attention in a recent editorial, *Science* 1993). Strangely enough, although Haldane correctly predicted that the nature and control of radiation would be better understood in the future, he seriously erred in misstating its commercial possibilities.

He wrote: 'I may add in parenthesis that, on thermodynamical grounds which I can hardly summarize shortly, I do not much believe in the commercial possibility of induced radio-activity.' Not only was Haldane later to witness the exploitation of atomic energy for both peaceful and offensive goals, he himself contributed to the estimation of genetic damage resulting from ionizing radiation. (Freeman Dyson informed me that Haldane was only echoing the prevailing sentiment among scientists in 1923; Dyson 1994, personal communication).

Another prediction of Haldane was concerned with the large-scale production of synthetic foodstuffs (p. 34): 'chemistry will be applied to the production of a still more important group of physiologically active substances, namely foods. . . . we may use micro-organisms, but in any case within the next century sugar and starch will be about as cheap as sawdust. Many of our foodstuffs, including the proteins, we shall probably build up from simpler sources such as coal and atmospheric nitrogen. I should be inclined to allow 120 years, but not much more, before a completely satisfactory diet can be produced in this way on a commercial scale.'

Haldane's *Daedalus* is repleat with quotable quotes. Commenting on two famous names, Haldane wrote (p. 26): 'At present physical theory is in a state of profound suspense. This is primarily due to Einstein—the greatest Jew since Jesus. I have no doubt that Einstein's name will still be remembered and revered when Lloyd George, Foch, and William Hohenzollern share with Charles Chaplin that inelecutable oblivion which awaits the uncreative mind. . . . I do not doubt that he will be believed. A prophet who can give signs in the heavens is always believed. No one ever seriously questioned Newton's theory after the return of Halley's comet.' On the subject of inter-planetary communication, Haldane wrote (p. 35): 'Whether this is possible I can form no conjecture; that it will be attempted I have no doubt whatever.'

With reference to important biological inventions, Haldane stated that four were invented before the dawn of history: the domestication of animals, the domestication of plants, the domestication of fungi for the production of alcohol, and another, a far reaching one from the viewpoint of sexual selection—women's face and breasts as the targets of a man's attention (a change from the steatopygous Hottentot to the modern European, as Haldane characterized it). He added two more, from the twentieth century—bactericide and the artificial control of conception (p. 35).

BRAVE NEW WORLD

Freeman Dyson categorically stated: 'Most of the biological inventions which Aldous Huxley used a few years later as background for his novel *Brave new world* were cribbed from Haldane's *Daedalus* (Dyson 1979). The same thought had occurred to me long before I read Professor Dyson's book. The Haldanes and the Huxleys knew each other intimately in Oxford. Both Julian and Aldous Huxley were frequent visitors to the Haldane household since their childhood. J.B.S. Haldane's parents—physiologist John Scott Haldane and his wife Louisa Kathleen—often showed much fondness for the Huxley children. Of special interest was the close friendship between Haldane's sister Naomi and Aldous Huxley (Mitchison 1985, personal communication). Both Aldous and Julian were closely familiar with Haldane's ideas and writings. These in turn inspired Aldous Huxley's *Brave new world*. What Huxley had achieved, however, was the popularization of Haldane's ideas on a large scale by incorporating them into a novel. Both works dealt with utopias of a special kind, those which are primarily based on advances in our biological knowledge.

HALDANE'S VIEWS ON EUGENICS: THE POST-*DAEDALUS* PERIOD

It is of interest to discuss briefly what changes, if any, were evident in J.B.S. Haldane's eugenic outlook during the post-*Daedalus* years of his life, and especially during his later years.

We might as well begin with a definition of 'eugenics'. Galton (1883) defined it as the 'science which deals with all influences that improve the inborn qualities of a race; also with those that develop them to the utmost advantage.' It is thus clear that Galton included both the inborn qualities as well as other influences which would significantly impact on their development under the title 'eugenics'. For this reason, Haldane was careful to take into account the interaction between various genetic and non-genetic factors in his discussions of eugenics (for instance, see Haldane 1933, 1934, 1938).

Haldane's analysis of nature–nurture interaction was first presented to a congress of philosophers (published in 1936), but was later expanded into a more thorough discussion of this subject (Haldane 1946). This may well have been the result of the controversies which were being generated by the suppression of Mendelian genetics by Trofim Lysenko in the Soviet Union during the 1940s. With respect to a sound eugenic policy, Haldane (1946) wrote: 'We are not justified in condemning a genotype absolutely unless we are sure that some other genotype exists which would excel it by all possible criteria in all possible environments. We can only be reasonably sure of this in the case of the grosser types of congenital and mental defect. A moderate degree of mental dullness may be a desideratum for certain types of monotonous but at present necessary work, even if in most or all existing nations there may turn out to be far too many people so qualified.' Writing on another aspect, Haldane (1934) pointed out that many eligible men are routinely eliminated during warfare. He wrote: 'Curiously enough, eugenic organizations rarely include a demand for peace in their programmes, in spite of the fact that modern war leads to the destruction of the fittest members of both sides engaged in it.'

It has become customary to compare the eugenic views of Haldane and H.J. Muller. For instance, Adams (1990) wrote: 'The Soviet A.S. Serebrovsky, the American H.J. Muller, and the Briton J.B.S. Haldane, three of the most distinguished geneticists of our century . . . exhibited a life long commitment to eugenic ideals.' It is a mistake to lump these three men together in this respect. The attitudes of Haldane and Muller towards eugenics were quite different. While Muller retained his enthusiasm for eugenic improvement throughout his life, Haldane became much more cautious in his later years. There are several reasons for Haldane's caution. It was due, in part, to his belief that our knowledge of human genetics is grossly inadequate. Another reason was his disillusionment with man's capacity to apply scientific developments wisely, which was the result of his experience in two world wars (see Lederberg 1968). Finally, Haldane (1964) believed that the existence of certain self-stabilizing properties of human populations ('genetic homeostasis' in the sense of Lerner 1954 and Neel 1958) might slow down the benefits of eugenic selection.

I believe that it is quite easy to make too much of Haldane's writings on eugenics. Haldane was a prolific writer on a great number of topics. He was given to a great deal of biological speculation and eugenics was a part of that process. His pronouncements on eugenic improvement were almost always biology based. Mazumdar's (1992) assessment of the source of Haldane's eugenical views agrees with my own evaluation.

From about 1930 onwards Haldane (along with R.A. Fisher and L. Hogben) led the way to establishing the foundations of human genetics on a firm scientific footing (see Dronamraju 1989, for references).* He profoundly disagreed with the views of those who were recommending wholesale sterilization of mentally defective individuals, petty criminals, and the chronic unemployed, the so-called 'social problem group' (Haldane 1934). During the heyday of his Marxist dalliance, Haldane was inclined to consider earlier eugenics in terms of a class struggle, as sterilization and other eugenic measures proposed would impact disproportionately on the economically disadvantaged. One unexpected result of the rigour of methodology of human genetics, which was established by the mathematical analyses of Hogben, Haldane, Fisher, and Penrose, was that it led to the destruction of class-bound eugenics.

In his Norman Lockyer lecture, entitled 'Human biology and politics', Haldane (1934) stated that although he would like to see a State medical service which would benefit the poor, the middle-class patient would be better off with a capitalist type of medical organization. In the same address, Haldane suggested preventive eugenic measures which could be implemented on the basis of the then newly acquired knowledge of human genetics, considering both dominant and recessive conditions. Haldane emphasized the multiplicity of causes for various types of congenital defects, including mental defects, and warned of the dangers of irresponsible sterilization of whole categories of people as it would open the door for government intervention with serious consequences.

Haldane then tackled the often-posed question of whether the mean innate intelligence of the population is declining: 'Men and women born into one economic class are constantly passing into a richer one if they possess more innate intelligence than the average of their class, into a poorer one if they possess less. But the poor breed faster than the rich. Hence the innately stupid breed faster than the innately clever, and the mean innate ability of the population is falling.' Haldane then stated that there are two good reasons to doubt this. First, he compared the populations of Moslems, Christians, and

* For a review of later developments in human genetics, see Dronamraju (1992) and McKusick (1992).

Jews in Islamic countries. During the last twelve centuries, followers of the prophet Mohammed who acquired great wealth have practised polygamy. On the other hand, Christians and Jews in those countries have practised monogamy. Hence, Haldane argued that Moslems would have acquired greater commercial ability than members of other religions and that a Turk would generally beat a Jew or an Armenian in a commercial deal. Since that is not the case Haldane concluded that if the rich in England bred faster than the poor the resulting population would not necessarily possess greater innate ability. The second point which Haldane emphasized is that the phrase 'innate ability' is meaningless. A may prove abler than B in a particular environment but not in all environments except in a few exceptional cases such as when B is a microcephalic idiot. In the same essay, Haldane (1934) predicted that application of the data of human biology to politics and ethics will probably be more complex than application of the data of physics to industry.

Haldane repeatedly returned to the possibilities of hypothetical biological intervention, but not practical eugenics in the sense of Huxley–Muller. In his contribution to a CIBA foundation symposium on the future of man, Haldane (1963*b*) speculated on the possibilities for human evolution in the next 10 000 years. He considered the method of deliberate inductions of mutations using chemical agents that are more specific than X-rays and some other agents. One method he described involved the incorporation of synthesized new genes into human chromosomes. Other methods include the duplication of existing genes to perpetuate the advantage of heterozygosity (hybrid vigour), and intranuclear grafting to enable our descendants to incorporate many valuable capacities of other species without losing their human capacities. For instance, the disease-resisting quality of many animal species could be incorporated without losing human consciousness and intelligence. Haldane cited gene grafting as a means to induce various desired phenotypes suited for special tasks. One of these tasks was special adaptation for long-distance space travel. The following comment was typical of Haldane's approach: 'A regressive mutation to the condition of our ancestors in the mid-pliocene, with prehensile feet, no appreciable heels, and an ape-like pelvis, would be still better' (Haldane 1963*b*). With reference to the unlikely prospect of encountering high gravitational fields, Haldane wrote (1936*b*): 'Presumably they should be short-legged or quadrupedal. I would back an achondroplasic against a normal man on Jupiter.'

In his address to the XI International Congress of Genetics, Haldane's (1964) eugenic predictions are tempered by his conviction that our knowledge of human genetics is grossly inadequate to warrant any serious eugenic planning. Indeed, among all the major biologists who discussed future eugenic possibilities, Haldane was unique in emphasizing the inadequacy of our technological knowledge. With respect to the possibilities over the next ten thousand years, he wrote (1964): 'It may take a thousand years or so before we have a knowledge of human genetics even as full as our present very incomplete knowledge of organic chemistry. Till then we can hardly hope to do much for evolution . . . If the capacity for consciousness and control of physiological processes is prized by posterity, steps will probably be taken to make it commoner, and it may be that ten thousand years hence our descendants will differ from us not only in achievements but in capacities and aspirations, to so great an extent that it is useless to attempt to follow them further.'

Haldane (1964) further wrote: 'Perhaps we could find an environment X where race A would prove superior to B, and an environment Y where B would prove superior to A. This is almost certainly the case for disease resistance. . . . There may well be similar inborn differences on the psychological level. . . . The appalling results of false beliefs on human genetics are exemplified in the recent history of Europe. Perhaps the most important thing which human geneticists can do for society at the moment is to emphasize how little they yet know. This is a thankless task. It is vastly easier to proclaim the equality or the inequality of different races as regards genetical endowment than to state, not merely that we are ignorant, but that insofar as the races may be adapted to different environments the question may be unanswerable.' Earlier, Haldane (1938) presented a sane and balanced account of the genetic basis of human polymorphism, which was especially intended to counteract the Nazi propaganda.

SCIENCE AND ETHICS

Haldane touched upon a great number of ethical issues in *Daedalus*. First, he included a whole range of ethical issues relating to the application of science to the conduct of wars in the twentieth century. Second, he raised a whole range of questions concerning the application of biological knowledge for the purpose of reproductive intervention. Even at that time, in his very first discussion of eugenics in *Daedalus*, Haldane appears to mock the kind of eugenical selection that has been (and still is) advocated by most followers of the eugenics movement, an attitude that is consistent with his later writings. Instead, Haldane's approach to the betterment of the human species was based on hypothetical technical advances in what was later called molecular genetics. He argued that the separation of sexual love and reproduction facilitates greater flexibility in manipulating the human genome. Haldane was well ahead of his time in broaching these biological and ethical issues. The term 'parent' acquires a new meaning in such a society. Directed mutation and control of *in vitro* fertilization would lead to a mass production of humans with specialized skills and talents. The genetic revolution which Haldane had anticipated is now in full swing, the only difference being that it is happening a few decades later than he predicted.

An underlying theme in *Daedalus* (and other writings of Haldane) was the fact that new scientific discoveries are constantly changing our ethical outlook. Ethics are bound by time, knowledge, and culture. New ethical duties arise from the application of new discoveries. One important aspect of his discussion dealt with the new ethical questions which are constantly arising from biological discoveries. There is a clear warning in *Daedalus* that progress in science must go hand in hand with progress in our ethical outlook.

Haldane returned to the subject of the impact of science on ethics repeatedly in his writings. In his book *The inequality of man and other essays* (1932*b*), Haldane listed five different ways in which science can impact on ethical situations.

'Science impinges upon ethics in at least five different ways. In the first place, by its application it creates new ethical situations. Two hundred years ago the news of a famine in China created no duty for Englishmen . . . To-day the telegraph and the steam-engine have made action possible, and it becomes an ethical problem what action, if any, is right. Secondly, it may create new duties by pointing out previously unexpected consequences of our actions . . . we may not all be of one mind as to whether a person likely to transmit club-foot or cataract to half his or her children should be compelled to abstain from parenthood.

Thirdly, science affects our whole ethical outlook by influencing our views as to the nature of the world—in fact, by supplanting mythology. One man may see men and animals as a great brotherhood of common ancestry . . . Another will regard even the noblest aspects of human nature as products of a ruthless struggle for existence . . . A third . . . will take refuge in a modified epicureanism . . . Fourthly . . . anthropology . . . is bound to have a profound effect . . . by showing that any given ethical code is only one of a number practiced with equal conviction and almost equal success; finally, ethics may be profoundly affected by an adoption of the scientific point of view; that is to say, the attitude which men of science, in their professional capacity, adopt towards the world. This attitude includes a high (perhaps an unduly high) regard for truth, and a refusal to come to unjustifiable conclusions . . . agnosticism.'

These thoughts are echoed in the foreword by Joshua Lederberg about our 'intergenerational responsibility'. In addition to the technological legacy, our children surely deserve adequate ethical and moral preparation to deal with new technologies.

In his essay, Dyson (p. 63) draws attention to the fact that both Haldane and Einstein shared common concerns in this regard. As Einstein later put it (1953): '. . . a positive aspiration and effort for an ethical–moral configuration of our common life is of overriding importance. Here no science can save us. I believe, indeed, that overemphasis on the purely intellectual attitude, often directed solely to the practical and factual, in our education, has led directly to the impairment of ethical values. I am not thinking so much of the dangers with which technical progress has directly confronted mankind, as of the stifling of mutual human considerations by a "matter-of-fact" habit of thought which has come to lie like a killing frost upon human relations.'

REFERENCES

Adams, M.B. (1990). *The wellborn science, eugenics in Germany, France, Brazil and Russia*. Oxford University Press, New York.

Clark, R.W. (1968). *JBS: the life and work of J.B.S. Haldane*. Hodder & Stoughton, London.

Clarke, A.C. (1968). Haldane and space. In *Haldane and modern biology* (ed. K.R. Dronamraju), pp. 243–8. Johns Hopkins University Press, Baltimore, MD.

Crow, J.F. (1984). The founders of population genetics. In *Human population genetics: the Pittsburgh symposium* (ed. A. Chakravarti), pp. 177–94. Van Nostrand Reinhold, New York.

Douglas, C.G., Haldane, J.S., and Haldane, J.B.S. (1912). The laws of combination of haemoglobin with carbon monoxide and oxygen. *Journal of Physiology*, **44**, 275–304.

Dronamraju, K.R. (ed.) (1968). *Haldane and modern biology*. Johns Hopkins University Press, Baltimore, MD.

Dronamraju, K.R. (1985). *Haldane: the life and work of J.B.S. Haldane with special reference to India*. Aberdeen University Press.

Dronamraju, K.R. (1987). On some aspects of the life and work of John Burdon Sanderson Haldane, F.R.S., in India. *Notes and Records of the Royal Society of London*, **41**, 211–37.

Dronamraju, K.R. (1989). *The foundations of human genetics*. Charles C. Thomas, Springfield.

Dronamraju, K.R. (ed.) (1990). *Selected genetic papers of J.B.S. Haldane*. Garland, London.

Dronamraju, K.R. (ed.) (1992). *The history and development of human genetics: progress in different countries*. World Scientific, London.

Dronamraju, K.R. (1993). *If I am to be remembered: the life and work of Julian Huxley with selected correspondence*. World Scientific, London.

Dyson, F. (1979). *Disturbing the universe*, p. 170. Harper & Row, New York.

Einstein, A. (1953). Letter read on the occasion of the seventy-fifth anniversary of the Ethical and Culture Society, New York, January 1951. *Mein Weltbild*. Europa Verlag, Zurich.

Galton, F. (1883). *Inquiries into human faculty and its development*. Macmillan, London.

Haldane, J.B.S. (1912). The dissociation of oxyhaemoglobin in human blood during partial CO poisoning. *Journal of Physiology*, **45**, 22.

Haldane, J.B.S. (1919). The combination of linkage values, and the calculation of distances between loci of linked factors. *Journal of Genetics*, **8**, 299–309.

Haldane, J.B.S. (1920). Some recent work on heredity. *Transactions of the Oxford University Junior Scientific Club*, **1**, 3–11.

Haldane, J.B.S. (1922). Sex ratio and unisexual sterility in hybrid animals. *Journal of Genetics*, **12**, 101–9.

Haldane, J.B.S. (1923). *Daedalus, or science and the future*. Kegan Paul, London.

Haldane, J.B.S. (1924*a*). *Daedalus, or science and the future*. E.P. Dutton, New York.

Haldane, J.B.S. (1924*b*). A mathematical theory of natural and artificial selection, part I. *Transactions of the Cambridge Philosophical Society*, **23**, 19–41.

Haldane, J.B.S. (1927). *Possible worlds and other essays*. Chatto & Windus, London.

Haldane, J.B.S. (1932*a*). *The causes of evolution*. Longmans, Green, London.

Haldane, J.B.S. (1932*b*). *The inequality of man and other essays*. Chatto & Windus, London.

Haldane, J.B.S. (1933). Biology and statesmanship. In *Biology in everyday life* (ed. J.R. Baker and J.B.S. Haldane). Allen and Unwin, London.

Haldane, J.B.S. (1934). *Human biology and politics: tenth annual Norman Lockyer lecture to the British Science Guild*. British Science Guild, London.

Haldane, J.B.S. (1936). Some principles of causal analysis in genetics. *Erkenntnis*, **6**, 346–56.

Haldane, J.B.S. (1937). Biochemistry of the individual. In *Perspectives in biochemistry* (ed. J. Needham and D.E. Green). Cambridge University Press.
Haldane, J.B.S. (1938). *Heredity and politics*. Allen & Unwin, London.
Haldane, J.B.S. (1946). The interaction of nature and nuture. *Annals of Eugenics*, **13**, 197–205.
Haldane, J.B.S. (1963*a*). Some lies about science. In *Rationalist Annual for 1963*; reprinted in *Science and life, essays of a rationalist*, Pemberton, London, 1968.
Haldane, J.B.S. (1963*b*). Biological possibilities for the human species in the next ten thousand years. In *Man and his future* (ed. G. Wolstenholme), pp. 337–61. J. & A. Churchill, London.
Haldane, J.B.S. (1964). The implications of genetics for human society. Proceedings of the 11th International Congress of Genetics. In *Genetics Today*, Vol. 2, (ed. S.J. Geerts), pp. xci–cii. Pergamon Press, Oxford.
Haldane, J.B.S. and Huxley, J.S. (1927). *Animal biology*. Clarendon Press, Oxford.
Haldane, J.B.S., Sprunt, A.D., and Haldane, N.M. (1915). Reduplication in mice (preliminary communication). *Journal of Genetics*, **5**, 133–5.
Huxley, A. (1923). *Antic hay*. Chatto & Windus, London.
Huxley, A. (1928). *Point counter point*. Chatto & Windus, London.
Huxley, A. (1932). *Brave new world*. Penguin, New York.
Huxley, J.S. (1920). Metamorphosis of axolotl caused by Thyroid-feeding. *Nature*, **104**, 435.
Huxley, J.S. (1970). *Memories*. Harper & Row, New York.
Lawrence, D.H. (1915). *The rainbow*. The Modern Library, New York.
Lawrence, D.H. (1928). *Lady Chatterley's lover*. Gieseppe Orioli, Florence (1st America edn. Grosset & Dunlap, New York, 1932).
Lederberg, J. (1968). Haldane's biology and social insight. In *Haldane and modern biology* (ed. K.R. Dronamraju), pp. 219–30. Johns Hopkins University Press, Baltimore, MD.
Lerner, I.M. (1954). *Genetic homeostasis*. Dover, New York.
McKusick, V.A. (1992). Human genetics: the last 35 years, the present, and the future. *American Journal of Human Genetics*, **50**, 663–70.
Mazumdar, P.M.H. (1992). *Eugenics, human genetics and human failings, the eugenics society, its source and its critics in Britain*. Routledge, London.
Mitchison, N. (1968). Beginnings. In *Haldane and modern biology* (ed. K.R. Dronamraju), pp. 299–305. Johns Hopkins University Press, Baltimore, MD.
Mitchison, N. (1985). Personal communication.
Neel, J.V. (1958). A study of major congenital defects in Japanese infants. *American Journal of Human Genetics*, **10**, 398–445.
Russell, B. (1924). *Icarus, or the future of science*. E.P. Dutton, New York.
Science (1993). Editorial. *Science*, **261**, 1255.
Seymour M. (1992). *Ottoline Morrell: life on the grand scale*. Farrar Straus Giroux, New York.
Weatherall, M. and Kamminga, H. (1992). *Dynamic science: biochemistry in Cambridge 1898–1949*. Wellcome Unit for the History of Medicine, Cambridge.
Wells, H.G. (1902). *Anticipations of the reaction of mechanical and scientific progress upon human life and thought*. Harper, London.

CHRONOLOGY OF J.B.S. HALDANE'S LIFE

1892 John Burdon Sanderson Haldane born, Oxford, Guy Fawkes Day, November 5.
1896 Sister Naomi (later Lady Mitchison) born.
1901 Attends a lecture by A.D. Darbishire on mendelian experiments.
1912 Publishes first scientific paper on human respiratory physiology (with his father J.S. Haldane and C.G. Douglas).
1914 Graduates with honours from Oxford University in literae humaniores (classics, philosophy, ancient history); world war I* starts, joins Third Battalion of the Black Watch.
1915 Sent to France and horrors of trench warfare; publishes first genetical paper on linkage in vertebrates (with A.D. Sprunt and sister Naomi, Sprunt killed in the war).
1917 Sent to India to recuperate from war injuries.
1919 Elected Fellow of New College, Oxford, in physiology; invents the first mapping function and centimorgan (cM) as a unit of distance in a chromosome.
1920 Formulates the gene–enzyme hypothesis.
1922 Proposes 'Haldane's rule': when in the F_1 offspring of two different animal races one sex is absent, rare, or sterile, that sex is the heterozygous sex.
1923 Appointed Sir William Dunn Reader in Biochemistry at Cambridge University; *Daedalus, or science and the future* (London).
1924 Publishes the first paper in his series 'mathematical theory of natural selection' (population genetics).
1925 Part-time head of the Genetical Department, John Innes Horticultural Institution in Merton; initiates research in the biochemical genetics of plant pigments; derives (with Briggs) the basic law of steady state kinetics for treating enzymatic catalysis.
1926 Marries Charlotte Burghes.
1927 *Possible worlds and other essays*; *Animal biology* (with Julian Huxley); carbon monoxide as a tissue poison (subject-investigator); estimates probability of gene fixation.
1929 Proposes his theory of the 'origin of life' in an anaerobic world.
1931 Evolution in 'metastable' populations.
1932 *The inequality of man and other essays*; *The causes of evolution* (proposes the indirect method for estimating the mutation rate of a human gene); elected Fellow of the Royal Society; Aldous Huxley's *Brave new world* published but makes no acknowledgement of *Daedalus*.
1933 Appointed professor of genetics at University College, London.
1935 *Science and the supernatural* (with A. Lunn); invents the formula for estimating mutation rates for sex-linked genes.
1936 Father, physiologist John Scott Haldane, dies.

* Haldane preferred to use lower case (w) when referring to world war because he used to say that he did not wish to glorify war. I follow the same practice.

1937 Effect of variation on fitness; first human gene map (with Julia Bell); appointed to Weldon Chair in Biometry at University College, London; resigns from the John Innes Horticultural Institution; biochemistry of the individual; joins the Spanish war effort.
1938 *Heredity and politics* (debunks Nazi theories of racist 'superiority').
1940 Experiments in diving physiology (subject-investigator); *Science in peace and war*.
1941 *New paths in genetics*.
1945 Divorces Charlotte Burghes; marries former pupil (and zoologist) Dr Helen Spurway.
1946 Interaction of nature and nurture.
1948 Croonian lecture, Royal Society (formal genetics of man).
1949 Proposes evolutionary basis of disease; suggests the term 'darwin' as a unit of evolutionary rate.
1954 *The biochemistry of genetics*; statics of evolution; measurement of natural selection.
1957 Moves to India; appointed Research Professor, Indian Statistical Institute, Calcutta; estimates 'cost of natural selection'.
1961 Awarded D.Sc. (Oxford) honoris causa; awarded Kimber medal by the US National Academy of Sciences; becomes citizen of India; mother, Louisa Kathleen (Trotter) Haldane, dies in Oxford.
1962 Moves to Bhubaneswar, Orissa (India) to found the Genetics and Biometry Laboratory.
1963 Elected foreign associate, US National Academy of Sciences.
1964 December 1, dies of cancer in Bhubaneswar, India; Haldane's body sent to the Medical College at Kakinada, south India, as desired in his will.
1976 Last posthumous publication *The man with two memories* (Merlin Press, London).
1977 Mrs. Haldane (Dr Helen Spurway) dies in Hyderabad, India.

CHAPTER TWO

DAEDALUS OR SCIENCE AND THE FUTURE

A PAPER READ TO THE HERETICS, CAMBRIDGE, ON FEBRUARY 4TH, 1923

J. B. S. HALDANE

SIR WILLIAM DUNN READER IN BIOCHEMISTRY, CAMBRIDGE UNIVERSITY

INTRODUCTION

I have slightly expanded certain parts of this paper since reading it. It has therefore probably lost any unity which it may once have possessed. It will be criticized for its undue and unpleasant emphasis on certain topics. This is necessary if people are to be induced to think about them, and it is the whole business of a university teacher to induce people to think.

DAEDALUS, OR SCIENCE AND THE FUTURE

As I sit down to write these pages I can see before me two scenes from my experience of the late war. The first is a glimpse of a forgotten battle of 1915. It has a curious suggestion of a rather bad cinema film. Through a blur of dust and fumes there appear, quite suddenly, great black and yellow masses of smoke which seem to be tearing up the surface of the earth and disintegrating the works of man with an almost visible hatred. These form the chief parts of the picture, but somewhere in the middle distance one can see a few irrelevant looking human figures, and soon there are fewer. It is hard to believe that these are the protagonists in the battle. One would rather choose those huge substantive oily black masses which are so much more conspicuous, and suppose that the men are in reality their servants, and playing an inglorious, subordinate, and fatal part in the combat. It is possible, after all, that this view is correct.

Had I been privileged to watch a battle three years later, the general aspect would have been very similar, but there would have been fewer men and more shell-bursts. There would probably, however, have been one very significant addition. The men would have been running, with mad terror in their eyes, from gigantic steel slugs, which were deliberately, relentlessly, and successfully pursuing them.

The other picture is of three Europeans in India looking at a great new star in the milky way. These were apparently all of the guests at a large dance who were interested in such matters. Amongst those who were at all competent to form views as to the origin of this cosmoclastic explosion, the most popular theory attributed it to a collision between two stars, or a star and a nebula. There seem, however, to be at least two possible alternatives to this hypothesis. Perhaps it was the last judgment of some inhabited world, perhaps a too successful experiment in induced radio-activity on the part of some of the dwellers there. And perhaps also these two hypotheses are identical, and what we were watching that evening was the detonation of a world on which too many men came out to look at the stars when they should have been dancing.

These two scenes suggest, very briefly, a part of the case against science. Has mankind released from the womb of matter a Demogorgon which is already beginning to turn against him, and may at any moment hurl him into the bottomless void? Or is Samuel Butler's even more horrible vision correct, in which man becomes a mere parasite of machinery, an appendage to the reproductive system of huge and complicated engines which will successively usurp his activities, and end by ousting him from the mastery of this planet? Is the machine-minder engaged on repetition-work the goal and ideal to which humanity is tending? Perhaps a survey of the present trend of science may throw some light on these questions.

But first we may consider for a moment, the question of whether there is any hope of stopping the progress of scientific research. It is after all a very recent form of human activity, and a sufficiently universal protest of mankind would be able to arrest it even now. In the middle ages public opinion made it so dangerous as to be practically impossible, and I am inclined to suspect that Mr. Chesterton, for example, would not be averse to a repetition of this state of things. The late M. Joseph Reinach, an able and not wholly illiberal thinker, publicly advocated it.

I think, however, that so long as our present economic and national systems continue, scientific research has little to fear. Capitalism, though it may not always give the scientific worker a living wage, will always protect him, as being one of the geese which produce golden eggs for its table. And competitive nationalism, even if war is wholly or largely prevented, will hardly forego the national advantages accruing from scientific research.

If we look at the other most probable alternative the prospect is little more hopeful. In this country the labour party alone among political organizations includes the fostering of research in its official programme. Indeed as far as biological research is concerned labour may prove a better master than capitalism, and there can be little doubt that it would be equally friendly to physical and chemical research if these came to lead immediately to shortened hours rather than to unemployment. In particular there is perhaps reason to think that that form of sentimentalism which hampers medical research in this country by legislation would be less likely to flourish in a robust and selfish labour party of the Australian type than in parties whose members enjoy the leisure which seems necessary to the development of such emotional luxuries.

It is of course possible that civilisation may collapse throughout the world as it has done in parts of Russia, and science with it, but such an event would in all probability, only postpone the problem for a few thousand years. And even in Russia we must not forget that first-rate scientific research is still being carried on.

The possibility has been suggested—I do not know how seriously—that the progress of science may cease through lack of new problems for investigation. Mr. Chesterton in *The Napoleon of Notting Hill*, a book written fifteen years or so ago, prophesied that hansom-cabs would still be in existence a hundred years hence owing to a cessation of invention. Within six years there was a hansom-cab in a museum, and now that romantic but tardy vehicle is a memory like the trireme, the velocipede, and the 1907 Voisin biplane. I do not suggest that Mr. Chesterton be dragged—a heavier Hector—behind the last hansom cab, but I do contend that, in so far as he claims to be a prophet rather than the voice of one crying in the wilderness, he may be regarded as negligible for the purpose of our discussion. I shall try shortly to show how far from complete are any branches of science at the present time.

But first a word on Mr. H. G. Wells might not be out of place. The very mention of the future suggests him. There are two points which I wish to make about Mr. Wells. In the first place, considered as a serious prophet, as opposed to a fantastic romancer, he is singularly modest. In 1902, for example, in a book called "Anticipations," he gave it as his personal opinion that by 1950 there would be heavier than air flying machines capable of practical use in war. That, said he, was his own view, though he was well aware that it would excite considerable ridicule. *I propose in this paper to make no prophecies rasher than the above.*

The second and more important point is that he is a generation behind the time. When his scientific ideas were formed, flying and radiotelegraphy, for example, were scientific problems, and the centre of scientific interest still lay in physics and chemistry. Now these are commercial problems, and I believe that the centre of scientific interest lies in biology. A generation hence it may be elsewhere, and the views expressed in this paper will appear as modest, conservative, and unimaginative as do many of those of Mr. Wells to-day.

I will only touch very briefly on the future of physics, as the subject is inevitably technical. At present physical theory is in a state of profound suspense. This is primarily due to Einstein—the greatest Jew since Jesus. I have no doubt that Einstein's name will still be remembered and revered when Lloyd George, Foch, and William Hohenzollern share with Charlie Chaplin that ineluctable oblivion which awaits the uncreative mind. I trust that I may be excused if I trespass from the strict subject of my theme to add my quota to the rather numerous mis-statements of Einstein's views which have appeared during the last few years.

Ever since the time of Berkeley it has been customary for the majority of metaphysicians to proclaim the ideality of Time, of Space, or of both. But they soon made it clear that in spite of this, time would continue to wait for no man, and space to separate lovers. The only practical consequence that they generally drew was that their own ethical and political views were somehow inherent in the structure of the universe. The experimental proof or disproof of such deductions is difficult, and—if the late war may be regarded as an experimental disproof of certain of Hegel's political tenets—costly and unsatisfactory.

Einstein, so far from deducing a new decalogue, has contented himself with deducing the consequences to space and time themselves of their ideality. These are mostly too small to be measurable, but some, such as the deflection of light by the sun's gravitational field, are susceptible of verification, and have been verified. The majority of scientific men are now being constrained by the evidence of these experiments to adopt a very extreme form of Kantian idealism. The Kantian *Ding-an-sich* is an eternal four-dimensional manifold, which we perceive as space and time, but what we regard as space and what as time is more or less fortuitous.

It is perhaps interesting to speculate on the practical consequences of Einstein's discovery. I do not doubt that he will be believed. A prophet who can give signs in the heavens is always believed. No one ever seriously questioned Newton's theory after the return of Halley's comet. Einstein has told us that space, time, and matter are shadows of the fifth dimension, and the heavens have declared his glory. In consequence Kantian idealism will become the basal working hypothesis of the physicist and finally of all educated men, just as materialism did after Newton's day. We may not call ourselves materialists, but we do interpret the activities of the moon, the Thames, influenza, and aeroplanes in terms of matter. Our ancestors did not, nor, in all probability, will our descendants. The materialism (whether conscious or sub-conscious does not very much matter) of the last few generations has led to various results of practical importance, such as sanitation, Marxian socialism, and the right of an accused person to give evidence on his or her own behalf. The reign of Kantian idealism as the basal working hypothesis, first of physics, and then of everyday life, will in all probability last for some centuries. At the end of that time a similar step in advance will be taken. Einstein showed that experience cannot be interpreted in terms of space and time. This was a well-known fact, but so long as space and time did not break down in their own special sphere, that of explaining the facts of motion, physicists continued to believe in them, or at any rate, what was much more important, to think in terms of them for practical purposes.

A time will however come (as I believe) when physiology will invade and destroy mathematical physics, as the latter have destroyed geometry. The basic metaphysical working hypothesis of science and practical life will then, I think, be something like Bergsonian activism. I do not for one moment suggest that this or any other metaphysical system has any claims whatever to finality.

Meanwhile we are in for a few centuries during which many practical activities will probably be conducted on a basis, not of materialism, but of Kantian idealism. How will this affect our manners, morals and politics? Frankly I do not know, though I think the effect will be as great as that of Newton's work, which created most of the intellectual forces of the 18th century. The Condorcets, Benthams, and Marxs of the future will I think be as ruthlessly critical of the metaphysics and ethics of their times as were their predecessors, but not quite so sure of their own; they will lack a certain heaviness of touch which we may note in Utilitarianism and Socialism. They will recognise that perhaps in ethics as in physics, there are so to speak fourth and fifth dimensions that show themselves by effects which, like the perturbations of the planet Mercury, are hard to detect even in one generation, but yet perhaps in the course of ages are quite as important as the three-dimensional phenomena.

If the quantum hypothesis is generally adopted even more radical alterations in our thinking will be necessary. But I feel it premature even to suggest their direction in the present unsatisfactory state of quantum mechanics. It may be that as Poincare (the other Poincare) suggested we shall be forced to conceive of all change as occurring in a series of clicks, and all space as consisting of discrete points. However this may be it is safe to say that a better knowledge of the properties of radiation will permit us to produce it in a more satisfactory manner than is at present possible. Almost all our present sources of light are hot bodies, 95% of whose radiation is invisible. To light a lamp as a source of light is about as wasteful of energy as to burn down one's house to roast one's pork. It is a fairly safe prophecy that in 50 years light will cost about a fiftieth of its present price, and there will be no more night in our cities. The alternation of day and night is a check on the freedom of human activity which must go the way of other spatial and temporal checks. In the long run I think that all that applied physics can do for us is to abolish these checks. It enables us to possess more, travel more, and communicate more. I shall not attempt to predict in detail the future developments of transport and communication. They are only limited by the velocity of light. We are working towards a condition when any two persons on earth will be able to be completely present to one another in not more than 1–24 of a second. We shall never reach it, but that is the limit which we shall approach indefinitely.

Developments in this direction are tending to bring mankind more and more together, to render life more and more complex, artificial, and rich in possibilities—to increase indefinitely man's powers for good and evil.

But there are two prerequisites for all progress of this kind, namely continuous supplies of human and mechanical power. As industries become more and more closely interwoven, so that a dislocation of any one will paralyse a dozen others (and that is the position towards which we are rapidly moving), the ideal of the leaders of industry, under no matter what economic system, will be directed less and less to the indefinite increase of production in the intervals between such dislocations, and more and more to stable and regular production, even at the cost of reduction of profits and output while the industry is proceeding normally. It is quite possible that capitalism itself may demand that the control of certain key industries be handed over completely to the workers in those industries, simply in order to reduce the number of sporadic strikes in them. And as industrial progress continues an ever larger number—perhaps the majority—of industries will become key industries. The solution may be entirely different—we may well see a return to feudalism. But the probability is that the problem will be solved. This view may seem optimistic, but it is more likely than the alternative thesis which may be briefly stated as follows: "*No human society will ever succeed in producing a stable organization in which the majority of the population is employed otherwise than in agriculture, animal-rearing, hunting or fishing.*" It took some thousands of years to produce the stable agricultural society which forms the basis of European life and whose morals we are too apt to regard as eternal truths. It should take a shorter time to evolve a stable industrial society. The people that do so will inherit the earth. In sum, I believe that the progress of science will ultimately make industrial injustice as self-destructive as it is now making international injustice.

As for the supplies of mechanical power, it is axiomatic that the exhaustion of our coal and oil-fields is a matter of centuries only. As it has often been assumed that their exhaustion would lead to the collapse of industrial civilization, I may perhaps be pardoned if I give some of the reasons which lead me to doubt this proposition.

Water-power is not, I think, a probable substitute, on account of its small quantity, seasonal fluctuation, and sporadic distribution. It may perhaps, however, shift the centre of industrial gravity to well-watered mountainous tracts such as the Himalayan foothills, British Columbia, and Armenia. Ultimately we shall have to tap those intermittent but inexhaustible sources of power, the wind and the sunlight. The problem is simply one of storing their energy in a form as convenient as coal or petrol. If a windmill in one's back garden could produce a hundredweight of coal daily (and it can produce its equivalent in energy), our coalmines would shut down to-morrow. Even to-morrow a cheap, foolproof, and durable storage battery may be invented, which will enable us to transform the intermittent energy of the wind into continuous electric power.

Personally, I think that four hundred years hence the power question in England may be solved somewhat as follows: The country will be covered with rows of metallic windmills working electric motors which in their turn supply current at a very high voltage to great electric mains. At suitable distances, there will be great power stations where during windy weather the surplus power will be used for the electrolytic decomposition of water into oxygen and hydrogen. These gases will be liquefied, and stored in vats, vacuum jacketed reservoirs, probably sunk in the ground. If these reservoirs are sufficiently large, the loss of liquid due to leakage inwards of heat will not be great; thus the proportion evaporating daily from a reservoir 100 yards square by 60 feet deep would not be 1-1000 of that lost from a tank measuring two feet each way. In times of calm, the gases will be recombined in explosion motors working dynamos which produce electrical energy once more, or more probably in oxidation cells. Liquid hydrogen is weight for weight the most efficient known method of storing energy, as it gives about three times as much heat per pound as petrol. On the other hand it is very light, and bulk for bulk has only one-third of the efficiency of petrol. This will not, however, detract from its use in aeroplanes, where weight is more important than bulk. These huge reservoirs of liquefied gases will enable wind energy to be stored, so that it can be expended for industry, transportation, heating, and lighting, as desired. The initial costs will be very considerable, but the running expenses less than those of our present system. Among its more obvious advantages will be the fact that energy will be as cheap in one part of the country as another, so that industry will be greatly decentralized; and that no smoke or ash will be produced.

It is on some such lines as these, I think, that the problem will be solved. It is essentially a practical problem, and the exhaustion of our coalfields will furnish the necessary stimulus for its solution. Even now perhaps Italy might achieve economic independence by the expenditure of a few million pounds upon research on the lines indicated. I may add in parenthesis that, on thermodynamical grounds which I can hardly summarize shortly, I do not much believe in the commercial possibility of induced radio-activity.

Before I turn to the principal part of my subject I should like to consider very briefly the influence on art and literature of our gradual conquest of space and time. I think that the blame for the decay of certain arts rests primarily on the defective education of the artists. An artist must understand his subject matter. At present not a single competent poet and very few painters and etchers outside the Glasgow School understand industrial life, and I believe that there is only one architect of any real originality who understands the possibilities of ferro-concrete. I do not know his name, but he produced in Soissons before the war a market-place with the dignity and daring of an ancient Egyptian temple. If I knew that he had been entrusted with the rebuilding of Soissons, I could not regret its destruction.

Now if we want poets to interpret physical science as Milton and Shelley did (Shelley and Keats were the last English poets who were at all up-to-date in their chemical knowledge), we must see that our possible poets are instructed, as their masters were, in science and economics. I am absolutely convinced that science is vastly more stimulating to the imagination than are the classics, but the products of this stimulus do not normally see the light because scientific men as a class are devoid of any perception of literary form. When they can express themselves we get a Butler or a Norman Douglas. Not until our poets are once more drawn from the educated classes (I speak as a scientist), will they appeal to the average man by showing him the beauty in his own life as Homer and Virgil appealed to the street urchins who scrawled their verses on the walls of Pompeii.

And if we must educate our poets and artists in science, we must educate our masters, labour and capital, in art. Personally I believe that we may have good hopes of both. The capitalist's idea of art in industry at present tends to limit itself to painting green and white stripes on the front of his factories in certain cases. This is a primitive type of decoration, but it has, I think, the root of the matter in it. Before long someone may discover that frescoes inside a factory increase the average efficiency of the worker 1.03% and art will become a commercial proposition once more. Even now it is being discovered that artistic advertising often pays. Similarly I do not doubt that labour will come to find that it cannot live by bread (or shall we say bread and beer) alone. But it can hardly be expected to make this discovery until it is assured of its supply of bread and beer.

Applied chemistry has introduced into human life no radical novelty of the importance of the heat-engine or the telegraph. It has vastly increased the production of various types of substance the most important being metals. But there were explosives, dyes, and drugs before chemistry was a science, and its progress along present lines will mainly alter life in a quantitative manner. Perhaps the biggest problems before it in metallurgy are the utilization of low-grade iron ores, and the production of aluminium from clay, which contains up to 24% of that metal. I do not think that even when this is accomplished aluminium will oust iron and steel as they ousted bronze and flint, but it and its alloys will certainly take the second, and possibly the first place as industrial metals. There is just a hope, though I fear little more, that a large-scale production of perfume may form the basis of a re-education of our rather rudimentary sense of smell, but the most interesting possibilities of chemical invention are very clearly in biological chemistry, and for the following reasons.

Desirable substances fall on the whole into two classes. The first are desirable on account of their physical or chemical properties, for example iron, wood or glass, which we use as a part of systems such as fires, houses, or razors, which procure us certain benefits. The second are desirable on account of their physiological properties. Such substances include foods, drinks, tobacco, and drugs. Colours and scents occupy an intermediate position. The value of this second class of substances rests on a quite special relationship to the human organism which depends in the most intimate way on the constitution of the latter, and has not in general been at all fully explained in terms of physics and chemistry. For example fires can be made of coal or peat instead of wood, but no other chemical substance has the same effect as water or alcohol. So unless a chemical substance has new physiological properties its production will merely serve to improve or make possible some appliance whose use lies within the sphere of applied physics. Within historical time two and only two substances of the second class have come into universal use in Europe, namely caffeine and nicotine, which were introduced into this country in the sixteenth and seventeenth centuries. There are others of immense importance, such as chloroform and quinine, but their use is not universal. But coffee, tea, and tobacco, with alcohol, are as much a part of normal life as food and water. There is no reason to suppose that the list of such substances is exhausted. During the war Embden* the professor of physiology in Frankfurt University discovered that a dose of about 7 grams of acid sodium phosphate increases a man's capacity for prolonged muscular work by about 20%, and probably aids in prolonged mental work. It can be taken over very lengthy periods. A group of coal-miners took it for nine months on end with very great effect on their output. It has no after-effects like those of alcohol, and one cannot take a serious over-dose as it merely acts as a purgative. (They gave certain Stosstruppen too much!) Thousands of people in Germany take it habitually. It is possible that it may become as normal a beverage as coffee or tea. It costs 1/9 per pound, or 1/3d. per dose.

The vast majority of chemical substances with physiological properties are unsuited for daily use like castor oil, or dangerous like morphine; probably none are without bad effects in certain cases. Those which are susceptible of daily use are of the utmost social importance. Tobacco has slight but definite effects on the character. Coffee-houses in London in the seventeenth and eighteenth centuries and cafes in modern Europe were and are civilizing influences of incalculable value. But these substances are profoundly obnoxious to a certain type of mind. It would perhaps be fantastic to suggest that Sir Walter Raleigh owed his death in part to his sovereign's objection to tobacco. But if he is not its proto-martyr it is at least probable that more men have died for tobacco smoking at the hands of Sikhs, Senussis, and Wahabis, whose religions forbid this practice, than died under the Roman empire for professing Christianity. Should it ever be generally realised that temperance is a mean we may expect that mankind will ultimately have at its disposal a vast array of substances like wine, coffee, and tobacco, whose intelligent use can

* Embden, Grafe, and Schmitz, *Zeitschrift für physiologische Chemie*, Vol. 113, p. 67, 1921.

add to the amenity of life and promote the expression of man's higher faculties.

But before that day comes chemistry will be applied to the production of a still more important group of physiologically active substances, namely foods. The facts about food are rather curious. Everyone knows that food is ultimately produced by plants, though we may get it at second or third hand if we eat animals or their products. But the average plant turns most of its sugar not into starch which is digestible, but into cellulose which is not, but forms its woody skeleton. The hoofed animals have dealt with this problem in their own way, by turning their bellies into vast hives of bacteria that attack cellulose, and on whose by-products they live. We have got to do the same, but outside our bodies. It may be done on chemical lines. Irvine has obtained a 95% yield of sugar from cellulose, but at a prohibitive cost. Or we may use micro-organisms, but in any case within the next century sugar and starch will be about as cheap as sawdust. Many of our foodstuffs, including the proteins, we shall probably build up from simpler sources such as coal and atmospheric nitrogen. I should be inclined to allow 120 years, but not much more, before a completely satisfactory diet can be produced in this way on a commercial scale.

This will mean that agriculture will become a luxury, and that mankind will be completely urbanized. Personally I do not regret the probable disappearance of the agricultural labourer in favour of the factory worker, who seems to me a higher type of person from most points of view. Human progress in historical time has been the progress of cities dragging a reluctant countryside in their wake. Synthetic food will substitute the flower garden and the factory for the dunghill and the slaughterhouse, and make the city at last self-sufficient.

There's many a strong farmer whose heart would break in two
If he could see the townland that we are riding to.
Boughs have their fruit and blossoms at all times of the year,
Rivers are running over with red beer and brown beer,
An old man plays the bagpipes in a golden and silver wood,
Queens, their eyes blue like the ice, are dancing in a crowd.

I should have liked had time allowed to have added my quota to the speculations which have been made with regard to inter-planetary communication. Whether this is possible I can form no conjecture; that it will be attempted I have no doubt whatever.

With regard to the application of biology to human life, the average prophet appears to content himself with considerable if rather rudimentary progress in medicine and surgery, some improvements in domestic plants and animals, and possibly the introduction of a little eugenics. The eugenic official, a compound, it would appear, of the policeman, the priest and the procurer, is to hale us off at suitable intervals to the local temple of Venus Genetrix with a partner chosen, one gathers, by something of the nature of a glorified medical board. To this prophecy I should reply that it proceeds from a type of mind as lacking in originality as in knowledge of human nature. Marriage "by numbers", so to speak, was a comparatively novel idea when proposed by Plato 2,300 years ago, but it has already actually been practised in various places, notably among the subjects of the Jesuits in Paraguay. It is moreover likely, as we shall see, that the ends proposed by the eugenist will be attained in a very different manner.

But before we proceed to prophecy I should like to turn back to the past and examine very briefly the half dozen or so important biological inventions which have already been made. By a biological invention I mean the establishment of a new relationship between man and other animals or plants, or between different human beings, provided that such relationship is one which comes primarily under the domain of biology rather than physics, psychology or ethics. Of the biological inventions of the past, four were made before the dawn of history. I refer to the domestication of animals, the domestication of plants, the domestication of fungi for the production of alcohol, and to a fourth invention, which I believe was of more ultimate and far-reaching importance than any of these, since it altered the path of sexual selection, focussed the attention of man as a lover upon woman's face and breasts, and changed our idea of beauty from the steatapygous Hottentot to the modern European, from the Venus of Brassempouy to the Venus of Milo. There are certain races which have not yet made this last invention. And in our own day two more have been made, namely bactericide and the artificial control of conception.

The first point that we may notice about these inventions is that they have all had a profound emotional and ethical effect. Of the four earlier there is not one which has not formed the basis of a religion. I do not know what strange god will have the hardihood to adopt

Charles Bradlaugh and Annie Besant in the place of Triptolemus and Noah, but one may remark that it is impossible to keep religion out of any discussion of the practices which they popularized.

The second point is perhaps harder to express. The chemical or physical inventor is always a Prometheus. There is no great invention, from fire to flying, which has not been hailed as an insult to some god. But if every physical and chemical invention is a blasphemy, every biological invention is a perversion. There is hardly one which, on first being brought to the notice of an observer from any nation which had not previously heard of their existence, would not appear to him as indecent and unnatural.

Consider so simple and time-honoured a process as the milking of a cow. The milk which should have been an intimate and almost sacramental bond between mother and child is elicited by the deft fingers of a milk-maid, and drunk, cooked, or even allowed to rot into cheese. We have only to imagine ourselves as drinking any of its other secretions, in order to realise the radical indecency of our relation to the cow.*

No less disgusting a priori is the process of corruption which yields our wine and beer. But in actual fact the processes of milking and of the making and drinking beer appear to us profoundly natural; they have even tended to develop a ritual of their own whose infraction nowadays has a certain air of impropriety. There is something slightly disgusting in the idea of milking a cow electrically or drinking beer out of tea-cups. And all this of course applies much more strongly to the sexual act.

I fancy that the sentimental interest attaching to Prometheus has unduly distracted our attention from the far more interesting figure of Daedalus. It is with infinite relief that amidst a welter of heroes armed with gorgon's heads or protected by Stygian baptisms the student of Greek mythology comes across the first modern man. Beginning as a realistic sculptor (he was the first to produce statues whose feet were separated) it was natural that he should proceed to the construction of an image of Aphrodite whose limbs were activ-

* The Hindus have recognized the special and physiological relation of man to the cow by making the latter animal holy. A good Hindu would no more kill a cow than his foster-mother. But the holiness of the cow has unfortunately extended to all its products, and the extensive use of cowdung in Indian religious ceremonies is disgusting to the average European. The latter however, is insensitive to the equally loathsome injunctions of the Catholic Church with regard to human marriage. It would perhaps be better if both marriage and milking could be secularized.

ated by quicksilver. After this his interest inevitably turned to biological problems, and it is safe to say that posterity has never equalled his only recorded success in experimental genetics. Had the housing and feeding of the Minotaur been less expensive it is probable that Daedalus would have anticipated Mendel. But Minos held that a labyrinth and an annual provision of 50 youths and 50 virgins were excessive as an endowment for research, and in order to escape from his ruthless economies Daedalus was forced to invent the art of flying. Minos pursued him to Sicily and was slain there. Save for his valuable invention of glue, little else is known of Daedalus. But it is most significant that, although he was responsible for the death of Zeus' son Minos he was neither smitten by a thunderbolt, chained to a rock, nor pursued by furies. Still less did any of the rather numerous visitors to Hades discover him either in Elysium or Tartarus. We can hardly imagine him as a member of the throng of shades who besieged Charon's ferry like sheep at a gap. He was the first to demonstrate that the scientific worker is not concerned with gods.

The unconscious mind of the early Greeks, who focussed in this amazing figure the dim traditions of Minoan science, was presumably aware of this fact. The most monstrous and unnatural action in all human legend was unpunished in this world or the next. Even the death of Icarus must have weighed lightly with a man who had already been banished from Athens for the murder of his nephew. But if he escaped the vengeance of the gods he has been exposed to the universal and agelong reprobation of a humanity to whom biological inventions are abhorrent, with one very significant exception. Socrates was proud to claim him as an ancestor.

The biological invention then tends to begin as a perversion and end as a ritual supported by unquestioned beliefs and prejudices. Even now surgical cleanliness is developing its rites and its dogmas, which, it may be remarked, are accepted most religiously by women. With the above facts in your minds I would ask you to excuse what at first sight might appear improbable or indecent in any speculations which appear below, and to dismiss from your minds the belief that biology will consist merely in physical and chemical discoveries as applied to men, animals and plants.

I say advisedly "will consist", for we are at present almost completely ignorant of biology, a fact which often escapes the notice of biologists, and renders them too presumptuous in their estimates of the present position of their science, too modest in their claims for its future. If for example we take a typical case of applied biology such as the detection and destruction of the cholera bacillus, we find a great deal of science involved, but the only purely biological principle is the very important but not very profound one that some bacteria kill some men. The really scientific parts of the process are the optical and chemical methods involved in the magnification, staining and killing of the bacilli. When on the other hand we come to immunization to typhoid we find certain purely biological principles involved which are neither simple nor at all completely understood.

Actually biological theory consists of some ancient but not very easily stated truths about organisms in general, due largely to Aristotle, Hippocrates and Harvey, a few great principles such as those formulated by Darwin, Mayer, Claude Bernard, and Mendel, and a vast mass of facts about individual organisms and their parts which are still awaiting adequate generalization.

Darwin's results are beginning to be appreciated, with alarming effects on certain types of religion, those of Weismann and Mendel will be digested in the course of the present century, and are going to affect political and philosophical theories almost equally profoundly. I need hardly say that these latter results deal with the question of reproduction and heredity. We may expect, moreover, as time goes on, that a series of shocks of the type of Darwinism will be given to established opinions on all sorts of subjects. One cannot suggest in detail what these shocks will be, but since the opinions on which they will impinge are deep-seated and irrational, they will come upon us and our descendants with the same air of presumption and indecency with which the view that we are descended from monkeys came to our grandfathers. But owing to man's fortunate capacity for thinking in watertight (or rather idea-tight) compartments, they will probably not have immediate and disruptive effects upon society any more than Darwinism had.

Far more profound will be the effect of the practical applications of biology. I believe that the progress of medicine has had almost, if not quite, as deep an effect on society in Western Europe as the industrial revolution. Apart from the important social consequences which have flowed from the partial substitution of the doctor for the priest, its net result has been that whereas four hundred years ago most people died in childhood, they now live on an average, (apart from the late war), until forty-five. Bad as our urban conditions often are, there is not a slum in the country which has a third of the infantile death-rate of the royal family in the middle ages. Largely as a result of this religion has come to lay less and less stress on a good death, and more and more on a good life, and its whole outlook has gradually changed in consequence. Death has receded so far into the background of our normal thoughts that when we came into somewhat close contact with it during the war most of us failed completely to take it seriously.

Similarly institutions which were based on short lives have almost wholly collapsed. For example the English land system postulated that the land-owner should die aged about forty, and be succeeded by his eldest son, aged about twenty. The son had spent most of his life on the estate, and had few interests outside it. He managed it at least as well as anyone else could have done. Nowadays the father dodders on till about eighty, and is generally incompetent for ten years before his death. His son succeeds him at the age of fifty or so, by which time he may be a fairly competent colonel or stockbroker, but cannot hope to learn the art of managing an estate. In consequence he either hands it over to an agent who is deprived of initiative and often corrupt, or runs it unscientifically, gets a low return, and ascribes to Bolshevism what he should really lay at the door of vaccination.

But to return, if I may use the expression, to the future, I am going to suggest a few obvious developments which seem probable in the present state of biological science, without assuming any great new generalizations of the type of Darwinism. I have the very best precedents for introducing a myth at this point, so perhaps I may be excused if I reproduce some extracts from an essay on the influence of biology on history during the 20th century which will (it is hoped) be read by a rather stupid undergraduate member of this university to his supervisor during his first term 150 years hence.

"As early as the first decade of the twentieth century we find a conscious attempt at the application of biology to politics in the so-called eugenic movement. A number of earnest persons, having discovered the existence of biology, attempted to apply it in its then very crude condition to the production of a race of super-men, and in certain countries managed to carry a good deal of legislation. They appear to have managed to prevent the transmission of a good deal of syphilis, insanity, and the like, and they certainly succeeded in producing the most violent opposition and hatred amongst the classes whom they somewhat gratuitously regarded as undesirable parents. (There was even a rebellion in Nebraska). However, they undoubtedly prepared public opinion for what was to come, and so far served a useful purpose. Far more important was the progress in medicine which practically abolished infectious diseases in those countries which were prepared to tolerate the requisite amount of state interference in private life, and finally, after the league's ordinance of 1958, all over the world; though owing to Hindu opposition, parts of India were still quite unhealthy up to 1980 or so.

But from a wider point of view the most important biological work in the first third of the century was in experimental zoology and botany. When we consider that in 1912 Morgan had located several Mendelian factors in the nucleus of Drosophila, and modified its sex-ratio, while Marmorek had taught a harmless bacillus to kill guinea-pigs, and finally in 1913 Brachet had grown rabbit embryos in serum for some days, it is remarkable how little the scientific workers of that time, and a fortiori the general public, seem to have foreseen the practical bearing of such results.

As a matter of fact it was not until 1940 that Selkovski invented the purple alga *Porphyrococcus fixator* which was to have so great an effect on the world's history. In the 50 years before this date the world's average wheat yield per hectar had been approximately doubled, partly by the application of various chemical manures, but most of all by the results of systematic crossing work with different races; there was however little prospect of further advance on any of these lines. *Porphyrococcus* is an enormously efficient nitrogen-fixer and will grow in almost any climate where there are water and traces of potash and phosphates in the soil, obtaining its nitrogen from the air. It has about the effect in four days that a crop of vetches would have had in a year. It could not, of course have been produced in the course of nature, as its immediate ancestors would only grow in artificial media and could not have survived outside a laboratory. Wherever nitrogen was the principal limiting factor to plant growth it doubled the yield of wheat, and quadrupled the value of grass land for grazing purposes. The enormous fall in food prices and the ruin of purely agricultural states was of course one of the chief causes of the disastrous events of 1943 and 1944. The food glut was also greatly accentuated when in 1942 the Q strain of *Porphyrococcus* escaped into the sea and multiplied with enormous rapidity. Indeed for two months the surface of the tropical Atlantic set to a jelly, with disastrous results to the weather of Europe. When certain of the plankton organisms developed ferments capable of digesting it the increase of the fish population of the seas was so great as to make fish the universal good that it is now, and to render even England self-supporting in respect of food. So great was the prosperity in England that in that year the coal-miner's union entered its first horse for the Derby (a horse-race which still took place annually at that time).

It was of course as the result of its invasion by *Porphyrococcus* that the sea assumed the intense purple colour which seems so natural to us, but which so distressed the more aesthetically minded of our great grand-parents who witnessed the change. It is certainly curious to us to read of the sea as having been green or blue. I need not detail the work of Ferguson and Rahmatullah who in 1957 produced the lichen which has bound the drifting sand of the world's deserts (for it was merely a continuation of that of Selkovski), nor yet the story of how the agricultural countries dealt with their unemployment by huge socialistic windpower schemes.

It was in 1951 that Dupont and Schwarz produced the first ectogenetic child. As early as 1901 Heape had transferred embryo rabbits from one female to another, in 1925 Haldane had grown embryonic rats in serum for ten days, but had failed to carry the process to its conclusion, and it was not till 1940 that Clark succeeded with the pig using Kehlmann's solution as medium. Dupont and Schwarz obtained a fresh ovary from a woman who was the victim of an aeroplane accident, and kept it living in their medium for five years. They obtained several eggs from it and fertilized them successfully, but the problem of the nutrition and support of the embryo was more difficult, and was only solved in the fourth year. Now that the technique is fully developed, we can take an ovary from a woman, and keep it growing in a suitable fluid for as long as twenty years, producing a fresh ovum each month, of which 90 per cent can be fertilized, and the embryos grown successfully for nine months, and then brought out into the air. Schwarz never got such good results, but the news of his first success caused an unprecedented sensation throughout the entire world, for the birthrate was already less than the deathrate in most civilised countries. France was the first country to adopt ectogenesis officially, and by 1968 was producing 60,000 children annually by this method. In most countries the opposition was far stronger, and was intensified by the Papal Bull "Nunquam prius audito", and the similar fetwa of the Khalif, both of which appeared in 1960.

As we know ectogenesis is now universal, and in this country less than 30 per cent of children are now born of woman. The effect on human psychology and social life of the separation of sexual love and reproduction which was begun in the 19th century and completed in the 20th is by no means wholly satisfactory. The old family life had certainly a good deal to commend it, and although nowadays we bring on lactation in women by injection of placentin as a routine, and thus conserve much of what was best in the former instinctive cycle, we must admit that in certain respects our great grandparents had the advantage of us. On the other hand it is generally admitted that the effects of selection have more than counterbalanced these evils. The small proportion of men and women who are selected as ancestors for the next generation are so undoubtedly superior to the average that the advance in each generation in any single respect, from the increased output of first-class music to the decreased convictions for theft, is very startling. Had it not been for ectogenesis there can be little doubt that civilization would have collapsed within a measurable time owing to the greater fertility of the less desirable members of the population in almost all countries.

It is perhaps fortunate that the process of becoming an ectogenetic mother of the next generation involves an operation which is somewhat unpleasant, though now no longer disfiguring or dangerous, and never physiologically injurious, and is therefore an honour but by no means a pleasure. Had this not been the case, it is perfectly possible that popular opposition would have proved too strong for the selectionist movement. As it was the opposition was very fierce, and characteristically enough this country only adopted its present rather stringent standard of selection a generation later than Germany, thought it is now perhaps more advanced than any other country in this respect. The advantages of thorough-going selection, have, however, proved to be enormous. The question of the ideal sex ratio is still a matter of violent discussion, but the modern reaction towards equality is certainly strong."

Our essayist would then perhaps go on to discuss some far more radical advances made about 1990, but I have only quoted his account of the earlier applications of biology. The second appears to me to be neither impossible nor improbable, but it has those features which we saw above to be characteristic of biological inventions. If reproduction is once completely separated from sexual love mankind will be free in an altogether new sense. At present the national character is changing slowly according to quite unknown laws. The problem of politics is to find institutions suitable to it. In the future perhaps it may be possible by selective breeding to change character as quickly as institutions. I can foresee the election placards of 300 years hence, if such quaint political methods survive, which is perhaps improbable, "Vote for Smith and more musicians", "Vote for O'Leary and more girls", or perhaps finally "Vote for Macpherson and a prehensile tail for your grandchildren". We can already alter animal species to an enormous extent, and it seems only a question of time before we shall be able to apply the same principles to our own.

I suggest then that biology will probably be applied on lines roughly resembling the above. There are perhaps equally great possibilities in the way of the direct improvement of the individual, as we come to know more of the physiological obstacles to the development of different faculties. But at present we can only guess at the nature of these obstacles, and the line of attack suggested in the myth is the one which seems most obvious to a Darwinian. We already know however that many of our spiritual faculties can only be manifested if certain glands, notably the thyroid and sex-glands, are functioning properly, and that very minute changes in such glands affect the character greatly. As our knowledge of this subject increases we may be able, for example, to control our passions by some more direct method than fasting and flagellation, to stimulate our imagination by some reagent with less after-effects than alcohol, to deal with perverted instincts by physiology rather than prison. Conversely there will inevitably arise possibilities of new vices similar to but even more profound than those opened up by the pharmacological discoveries of the 19th century.

The recent history of medicine is as follows. Until about 1870 medicine was largely founded on physiology, or, as the Scotch called it "Institutes of medicine". Disease was looked at from the point of view of the patient, as injuries still are. Pasteur's discovery of the nature of infectious disease transformed the whole outlook, and made it possible to abolish one group of diseases. But it also diverted scientific medicine from its former path, and it is probable that, were bacterial unknown, though many more people would die of sepsis and typhoid, we should be better able to cope with kidney disease and cancer. Certain diseases such as cancer are probably not due to specific organisms, whilst others such as phthisis are due to forms which are fairly harmless to the average person, but attack others for unknown reasons. We are not likely to deal with them effectually on Pasteur's lines, we must divert our view from the micro-organism to the patient. Where the doctor cannot deal with the former he can often keep the patient alive long enough to be able to do so himself. And here he has to rely largely on a knowledge of physiology. I do not say that a physiologist will discover how to prevent cancer. Pasteur started life as a crystallographer. But whoever does so is likely at least to make use of physiological data on a large scale.

The abolition of disease will make death a physiological event like sleep. A generation that has lived together will die together. I suspect that man's desire for a future life is largely due to two causes, a feeling that most lives are incomplete, and a desire to meet friends from whom we have parted prematurely. A gentle decline into the grave at the end of a completed life's work will largely do away with the first, and our contemporaries will rarely leave us sorrowing for long.

Old age is perhaps harder on women than on men. They live longer, but their life is too often marred by the sudden change which generally overtakes them between forty and fifty, and sometimes leaves them a prey to disease, though it may improve their health. This change seems to be due to a sudden failure of a definite chemical substance produced by the ovary. When we can isolate and synthesize this body we shall be able to prolong a woman's youth, and allow her to age as gradually as the average man.

Psychology is hardly a science yet. Like biology it has arrived at certain generalizations of a rather abstract and philosophic character, but these are still to some extent matters of controversy. And though a vast number of most important empirical facts are known, only a few great generalizations from them—such as the existence of the subconscious mind—have yet been made. But anyone who has seen even a single example of the power of hypnotism and suggestion must realise that the face of the world and the possibilities of existence will be totally altered when we can control their effects and standardize their application, as has been possible, for example, with drugs which were once regarded as equally magical. Infinitely greater, of course, would be the results of the opening up of systematic communication with spiritual beings in another world, which is claimed as a scientific possibility. Spiritualism is already Christianity's most formidable enemy, and we have no data which allows us to estimate the probable effect on man of a religion whose dogmas are a matter of experiment, whose mysteries are prosaic as electric lighting, whose ethics are based on the observed results in the next world of a good or bad life in this. Yet that is the prospect before us if spiritualism obtains the scientific verification which it is now demanding, not perhaps with great success.

I have only been able, in the time at my disposal, to traverse a very few of the possible fields of scientific advance. If I have convinced anyone present that science has still a good deal up her sleeve, and that of a sufficiently startling character, I shall be amply repaid. If anything I have said appears to be of a gratuitously disgusting nature, I would reply that certain phenomena of normal life do seem to many to be of that nature, and that these phenomena are of the utmost scientific and practical importance.

I have tried to show why I believe that the biologist is the most romantic figure on earth at the present day. At first sight he seems to be just a poor little scrubby underpaid man, groping blindly amid the mazes of the ultramicroscopic, engaging in bitter and lifelong quarrels over the nephridia of flatworms, waking perhaps one morning to find that someone whose name he has never heard has demolished by a few crucial experiments the work which he had hoped would render him immortal. There is real tragedy in his life, but he knows that he has a responsibility which he dare not disclaim, and he is urged on, apart from all utilitarian considerations, by something or someone which he feels to be higher than himself.

The conservative has but little to fear from the man whose reason is the servant of his passions, but let him beware of him in whom reason has become the greatest and most terrible of the passions. These are the wreckers of outworn empires and civilisations, doubters, disintegrators, deicides. In the past they have been, in general, men like Voltaire, Bentham, Thales, Marx, and very possibly the divine Julius, but I think that Darwin furnishes an example of the same relentlessness of reason in the field of science. I suspect that as it becomes clear that at present reason not only has a freer play in science than elsewhere, but can produce as great effects on the world through science as through politics, philosophy, or literature, there will be more Darwins. Such men are interested primarily in truth as such, but they can hardly be quite uninterested in what will happen when they throw down their dragon's teeth into the world.

I do not say that biologists as a general rule try to imagine in any detail the future applications of their science. The central problems of life for them may be the relationship between the echinoderms and brachiopods, and the attempt to live on their salaries. They do not see themselves as sinister and revolutionary figures. They have no time to dream. But I suspect that more of them dream than would care to confess it.

I have given above a very small selection from my dreams. Perhaps they are bad dreams. It is of course almost hopeless to attempt any very exact prophecies as to how in detail scientific knowledge is going to revolutionize human life, but I believe that it will continue to do so, and even more profoundly than I have suggested. And though personally I am Victorian enough in my sympathies to hope that after all family life, for example, may be spared, I can only reiterate that not one of the practical advances which I have predicted is not already fore-shadowed by recent scientific work. If a chemist or physicist living at the end of the seventeenth century had been asked to predict the future application of his science he would doubtless have made many laughable errors in the best Laputan style, but he would have been certain that it would somehow be applied, and his faith would have been justified.

We must regard science then from three points of view. First it is the free activity of man's divine faculties of reason and imagination. Secondly it is the answer of the few to the demands of the many for wealth, comfort and victory, for "*νόσων τ' ἀπείρους καὶ μακραίωνας βίους*," gifts which it will grant only in exchange for peace, security and stagnation. Finally it is man's gradual conquest, first of space and time, then of matter as such, then of his own body and those of other living beings, and finally the subjugation of the dark and evil elements in his own soul.

None of these conquests will ever be complete, but all, I believe will be progressive. The question of what he will do with these powers is essentially a question for religion and aesthetic. It may be urged that they are only fit to be placed in the hands of a being who has learned to control himself, and that man armed with science is like a baby with a box of matches.

The answer to this contention may, I think, be found in the daily papers. For scores of centuries idealists had urged that wars must cease and all the earth be united under one rule. As long as any other alternative was possible it was persisted in. The events of the last nine years constituted a reductio ad absurdum of war, but when we ask who was responsible for this we shall find that it was not the visionaries but men like Black, Kekule, and Langley, who enlarged man's power over nature until he was forced by the inexorable logic of facts to form the nucleus of an international government.

We have already reacted against the frame of mind that engendered the league of nations, but we have not reacted at all completely. The league exists and is working, and in every country on earth there are many people, and ordinary normal people, who favour the idea in one form or another of a world state. I do not suggest that a world-state will arise from the present league—or for the matter of that from the third international. I merely observe that there is a wide-spread and organized desire for such an institution, and several possible nuclei for it. It may take another world-war or two to convert the majority. The prospect of the next world-war has at least this satisfactory element. In the late war the most rabid nationalists were to be found well behind the front line. In the next war no one will be behind the front line. It will be brought home to all whom it may concern that war is a very dirty business.

No doubt there is a fair chance that the possibility of human organization on a planetary scale may be rendered impossible by such a war. If so mankind will probably have to wait for a couple of thousand years for another opportunity. But to the student of geology such a period is negligible. It took man 250,000 years to transcend the hunting pack. It will not take him so long to transcend the nation.

I think then that the tendency of applied science is to magnify injustices until they become too intolerable to be borne, and the average man whom all the prophets and poets could not move, turns at last and extinguishes the evil at its source. Marx' theory of industrial evolution is a particular example of this tendency, though it does not in the least follow that his somewhat artificial solution of the problem will be adopted.

It is probable that biological progress will prove to be as incompatible with certain of our social evils as industrial progress has proved to be with war or certain systems of private ownership. To take a concrete example it is clear that the second biological invention considered by my future essayist would be intolerable in conjunction with our present system of relations between classes and sexes. Moral progress is so difficult that I think any developments are to be welcomed which present it as the naked alternative to destruction, no matter how horrible may be the stimulus which is necessary before man will take the moral step in question.

To sum up then, science is as yet in its infancy, and we can foretell little of the future save that the thing that has not been is the thing that shall be; that no beliefs, no values, no institutions are safe. So far from being an isolated phenomenon the late war is only an example of the disruptive results that we may constantly expect from the progress of science. The future will be no primrose path. It will have its own problems. Some will be the secular problems of the past, giant flowers of evil blossoming at last to their own destruction. Others will be wholly new. Whether in the end man will survive his accessions of power we cannot tell. But the problem is no new one. It is the old paradox of freedom re-enacted with mankind for actor and the earth for stage. To those who believe in the divinity of that part of man which aspires after knowledge for its own sake, who are able, in the words of Boethius:

> "te cernere finis
> "Principium, vector, dux, semita, terminus idem".

the prospect will appear most hopeful. But it is only hopeful if mankind can adjust its morality to its powers. If we can succeed in this, then science holds in her hands one at least of the keys to the thorny and arduous path of moral progress, then:

> "Per cruciamina leti
> "Via panditur ardua justis,
> "Et ad astra doloribus itur".

That is possibly a correct large-scale view, but it is only for short periods that one can take views of history sufficiently broad to render the fate of one's own generation irrelevant. The scientific worker is brought up with the moral values of his neighbours. He is perhaps fortunate if he does not realize that it is his destiny to turn good into evil. The moral and physical (though not the intellectual) virtues are means between two extremes. They are essentially quantitative. It follows that an alteration in the scale of human power will render actions bad which were formerly good. Our increased knowledge of hygiene has transformed resignation and inaction in face of epidemic disease from a religious virtue to a justly punishable offence. We have improved our armaments, and patriotism, which was once a flame upon the altar, has become a world-devouring conflagration.

The time has gone by when a Huxley could believe that while science might indeed remould traditional mythology, traditional morals were impregnable and sacrosanct to it. We must learn not to take traditional morals too seriously. And it is just because even the least dogmatic of religions tends to associate itself with some kind of unalterable moral tradition, that there can be no truce between science and religion.

There does not seem to be any particular reason why a religion should not arise with an ethic as fluid as Hindu mythology, but it has not yet arisen. Christianity has probably the most flexible morals of any religion, because Jesus left no code of law behind him like Moses or Muhammad, and his moral percepts are so different from those of ordinary life that no society has ever made any serious attempt to carry them out, such as was possible in the case of Israel and Islam. But every Christian church has tried to impose a code of morals of some kind for which it has claimed divine sanction. As these codes have always been opposed to those of the gospels a loophole has been left for moral progress such as hardly exists in other religions. This is no doubt an argument for Christianity as against other religions, but not as against none at all, or as against a religion which will frankly admit that its mythology and morals are provisional. That is the only sort of religion that would satisfy the scientific mind, and it is very doubtful whether it could properly be called a religion at all.

No doubt many people hope that such a religion may develop from christianity. The human intellect is feeble, and there are times when it does not assert the infinity of its claims. But even then:

> "Though in black jest it bows and nods
>
> *   *   *   *   *
>
> "I know it is roaring at the gods
> Waiting the last eclipse."

The scientific worker of the future will more and more resemble the lonely figure of Daedalus as he becomes conscious of his ghastly mission, and proud of it.

> "Black is his robe from top to toe,
> His flesh is white and warm below,
> All through his silent veins flow free
> Hunger and thirst and venery,

But in his eyes a still small flame
Like the first cell from which he came
Burns round and luminous, as he rides
Singing my song of deicides."

REVIEW OF *DAEDALUS, OR SCIENCE AND THE FUTURE*, REPRODUCED WITH PERMISSION FROM *NATURE*, 113, 1924, p. 740.

Mr Haldane has been known for some time to his acquaintances as a man of encyclopaedic knowledge and portentous versatility, as well as of high scientific attainment; but this little book is his first public venture outside the covers of scientific journals. It is an amplification of a paper read before the Heretics Club in Cambridge; even allowing for the expansion, the paper must have provided a 'crowded hour of glorious life'!

Mr Haldane sets out to explain that the future of science, in his opinion, will lie more and more with biology, just as its past has lain predominantly in the fields of physics and chemistry. He then goes on to point out that so far, almost every invention of practical utility has been physico-chemical—that is, has given man increased power without in any way affecting or coming to close quarters with his nature as an organism. Of important biological inventions there have been a bare half-dozen, and four of these are older than history. 'Bactericide' and the artificial control of conception are the only two modern ones the claims of which he is ready to admit. But clearly, he continues (and clearly too he is right), we are on the verge of scores of new inventions of a biological nature. The control of the sex-ratio is a matter of a few decades; the applications of endocrinology are beginning already; and we shall doubtless soon be adding considerably to the list of substances which, like alcohol or tea or tobacco, have such pleasant or stimulating properties that they will come into general use.

The author's further prophecies are cast in amusing form. He supposes a Cambridge undergraduate in AD 2024 reading an essay to his tutor on the progress of applied biology in the past century. We are introduced to artificially evolved Nitrogenfixing organisms, to the artificial alteration of character through applied endocrinology, and finally to 'ectogenesis'—the artificial development of human embryos outside the body, from ova taken from ovaries cultivated *in vitro* and artificially fertilised.

Far-fetched, this last prophecy? Not so very, if what has already been done with tissue-culture is remembered. As Mr Haldane says, think of the rapidity at which human evolution will then be able to progress, once the whole of each generation can be reared from a few dozen parents of selected type.

There are some biological possibilities which Mr Haldane seems to have forgotten—what, for example, of the regulation of sleep? There is surely room for improvement there. What, too, of the difficulties which civilisation seems always to bring in its trail? How will biology prevail upon men to be moderate and to live healthy lives? Half our present diseases are diseases of excess or of carelessness.

However, one cannot prophesy everything. We hope that Mr Haldane's booklet will not lack readers—they will find in it not only entertainment but also food for much thought.

Anonymous reviewer

CHAPTER THREE

J.B.S. HALDANE

M.F. PERUTZ

Editor's note. The following appreciation of Haldane was initially written by M.F. Perutz for the J.B.S. Haldane centenary celebration in India. It is included in this book with his kind permission. During the years 1923–33, Haldane held the position of Sir William Dunn Readership in Biochemistry at Cambridge University and was professionally identified as a 'biochemist'. It was during those years that Haldane conducted research in enzyme kinetics and directed research on the biochemical genetics of plant pigments. *Daedalus* was published in 1923, soon after he moved from Oxford to take up his new position at Cambridge.

In the 1920s and 1930s, Cambridge became one of the world centres of biochemistry. The subject was dominated by three outstanding men: Frederick Gowland Hopkins, David Keilin, and J.B.S. Haldane. Hopkins had to struggle for his then revolutionary idea that in 'the living cell . . . substances identifiable by chemical methods undergo changes which can be followed by chemical methods. . . . These substances are activated, and their reactions directed in space and time by . . . intracellular enzymes' (Needham and Baldwin 1937). Vitalism was still rife in those days, and even today some philosophers reject such ideas as Hopkins' as reductionism.

Keilin was a parasitologist turned biochemist who in 1925 discovered the vital cytochrome system of enzymes in the visible absorption spectrum of the muscle of a fly.

J.B.S. Haldane forged the link between enzymology and genetics. He was the first to point out that many enzymatic reactions as well as the blood groups are genetically controlled. He was also the first to suggest that the gene reproduces itself by a template mechanism via an intermediate 'negative'. 'Two possibilities are now open. The gene is a catalyst making a particular antigen, or the antigen is simply the gene or part of it let loose from its connexion with the chromosome. The gene has two properties. It intervenes in metabolism, sometimes at least by making a definite substance. And it reproduces itself. The gene, considered as a molecule must be spread out in a layer one *Baustein* deep. Otherwise it could not be copied. The most likely method of copying is by a process analogous to crystallization, a second similar layer of *Baustein* being laid down on the first. But we could conceive of a process analogous to the copying of a gramophone record by the intermediation of a "negative" perhaps related to the original as an antibody to an antigen' (Haldane 1937, pp. 8–9).

Like nearly all his contemporaries, he dismissed the idea that genes might be made of nucleic acids and asserted that they are most likely to be histones.

On the other hand, he had the right concept of enzymatic catalysis when he wrote: 'Using Fischer's lock and key simile, the key does not fit the lock quite perfectly, but exercises a certain strain on it', thus foreshadowing Linus Pauling's prediction in 1948 that the active sites of enzymes are likely to be complementary to the transition states of their substrates.

In 1950 Haldane took a keen interest in my early work on sickle cell haemoglobin, probably because he had first suggested in the previous year that individuals with various red cell disorders might be protected from fatal infections with malaria (Haldane 1949). He also predicted that the high incidence of thalassaemia in malarial countries might have arisen by balanced polymorphism, heterozygotes for the disease having a selective advantage relative to homozygotes of either kind.

I have a postcard from Haldane, dated 19 March 1954, where he writes 'Congratulations on your election to the Royal Society. I hope you will get some haemoglobin from Horlein and Weber's methaemoglobinaemia family to see if it is as like normal as the sickle cell stuff.' This refers to the first discovery of an abnormal haemoglobin in 1948, the year before the publication of Pauling *et al.* (1949). Horlein and Weber (1948) investigated a German family that exhibited hereditary cyanosis, believed to be due to a congenital heart defect. They found their hearts to be healthy, but their red cells contained a large fraction of methaemoglobin, owing to an unknown defect in the globin. Horlein and Weber's (1948) work appeared in the *Deutsche Medizinische Wochenschrift*, a little read journal in those early postwar years. It was characteristic of Haldane that he knew of this paper which was recognized much later as a milestone in medical genetics.

I did not get any blood from Horlein and Weber's family until nearly 20 years later when the chemical defect had been recognized as a replacement of the distal histidine in the haem pockets of the β-chains by tyrosine, but this haemoglobin failed to crystallize and I was never able to determine its structure and therefore could not answer Haldane's pertinent question.

REFERENCES

Haldane, J.B.S. (1937). The biochemistry of the individual. In *Perspectives in biochemistry* (ed. J. Needham and D.E. Green), pp. 1–10. Cambridge University Press.

Haldane, J.B.S. (1949). Disease and evolution. *La Ricerca Scientifica*, **19** (suppl.), 3–11.

Horlein, H. and Weber, G. (1948). Uber chronische familiare methamoglobinamie und eine neue modifikation des methamoglobins. *Deutsche Medizinische Wochenschrift*, 73, 476–8.

Needham, J. and Baldwin, E. (eds.) (1937). *Hopkins and biochemistry*. Heffers, Cambridge.

Pauling, L. Itano, H.A., Singer, S.J., and Wells, I.C. (1949). Sickle cell anemia, a molecular disease. *Science*, **110**, 543–8.

CHAPTER FOUR

Daedalus after Seventy Years

FREEMAN DYSON

My copy of *Daedalus* once belonged to Einstein. Unfortunately he left no written record of his response to it. The only evidence that he read it is a pencil-mark in the margin, presumably marking a passage that resonated with his own thinking. It appears beside the sentence (p. 28), in which Haldane discusses the long-range effect of the theory of relativity on the conceptual basis of ethics: '. . . in ethics as in physics, there are so to speak fourth and fifth dimensions that show themselves by effects which, like the perturbations of the planet Mercury, are hard to detect even in one generation, but yet perhaps in the course of ages are quite as important as the three dimensional phenomena.' Einstein's theory describes space–time as a continuum with four dimensions. Haldane's mention of a fifth dimension refers to a five-dimensional version of general relativity which had been invented by Theodor Kaluza in 1921. It is the remote ancestor of the superstring theories that are fashionable seventy years later.

We do not know when Einstein acquired his copy of *Daedalus* or when he read it. It is unlikely that he bought it, since he was not fluent enough in English to read it for pleasure. Most probably, the book was given to him soon after it was published in 1924, either by the author or by the publisher. Haldane revered Einstein as he revered nobody else, expressing his reverence with the memorable words that John Reed had applied less appropriately to Leon Trotsky: 'The greatest Jew since Jesus'. It is consistent with what we know about Einstein's way of thinking, that he believed a fundamental reform of concepts to be as necessary in ethics as in physics. If we can trust the evidence of the pencil-mark, Einstein believed this already in the 1920s. He evidently grasped at once the main message of Haldane's book, the message that the progress of science is destined to bring enormous confusion and misery to mankind unless it is accompanied by progress in ethics. This message was unwelcome to the scientists of Haldane's time and is equally unwelcome today.

Haldane also said, 'We are in for a few centuries during which many practical activities will probably be conducted on a basis, not of materialism, but of Kantian idealism.' This prediction turned out to be wildly wrong, so far as the majority of ordinary people living in the twentieth century are concerned. Few of us know what Kantian idealism means, and even fewer look to Kantian idealism to guide us in the conduct of our practical activities. Einstein was one of the few. Einstein was at home in a Kantian universe. For him, Haldane's prediction made sense. Perhaps, after another hundred years, it will make sense to more of us.

The pages devoted to Kantian idealism are the weakest part of Haldane's essay. Everywhere else, his shots are right on target. He did not know much about theoretical physics, but he had expert first-hand knowledge of warfare and physiology, the two subjects that provide the core of his argument. His knowledge of war and physiology was not theoretical but practical, gained by risking his life repeatedly in the trenches of France and the coal-mines of Staffordshire. Everything he says about warfare, and everything he says about physiology and medicine, ring true. He began his education in physiology by serving as his father's assistant in investigations of the effects of noxious gases to which miners and sailors were exposed in coal-mines and submarines. Later, he took part in similar investigations of the effects of more lethal gases on soldiers in the First World War. He fought with legendary bravery as an infantry officer on the Western Front. When he writes in *Daedalus*, 'Death has receded so far into the background of our normal thoughts that when we came into somewhat close contact with it during the war most of us failed completely to take it seriously,' we can be sure that he is recording his personal experience. *Daedalus* begins with an artillery bombardment on the Western Front, the shell-bursts nonchalantly annihilating the human protagonists who are supposed to be in charge of the battle. This opening scene epitomizes Haldane's hard-headed view of war. And likewise at the end, when the biologist in his laboratory, 'just a poor little scrubby underpaid man groping blinding amid the mazes of the ultramicroscopic', is transfigured into the mythical figure of Daedalus, 'conscious of his ghastly mission and proud of it', this closing scene epitomizes Haldane's hard-headed view of science. Haldane was saying, seventy years ago, that the destiny of the scientist is to turn good into evil, that the horrors of the First World War are not an isolated phenomenon but only an example of the disruptive consequences that we may constantly expect to emerge from the progress of science. Now, seventy years later, we are beginning to see more clearly what he had in mind.

The recent decision of the United States Congress to kill the Superconducting Supercollider project was a clear demonstration of the present unpopularity of science. The decision came as a severe shock to many of my scientist colleagues, but it would not have surprised Haldane. At the time when Haldane was writing in the 1920s, science was intensely unpopular in England because it was identified in the public mind with the technological carnage of the recent war. The First World War was seen as peculiarly evil, because the organizers and promoters of the slaughter were old men while the victims were young. The public blamed scientists in general, and chemists in particular, for the invention of explosives and poison gases which killed or scarred a whole generation of young Englishmen. Scientists were seen as a privileged priesthood, callously profiting from the misery of the unprivileged. Forty years later in America, a similar hatred of science was aroused in the generation of young people who experienced the consequences of technology in the Vietnam war and felt themselves to be victims. And now, seventy years after Haldane predicted it, science has once again turned good into evil. This time the evil is not a war, but a civilian technology that systematically widens the gulf between rich and poor, deprives uneducated young people of jobs, and leaves large numbers of young mothers and children homeless and hopeless. When one walks through the streets of New York after dark during the Christmas season, one sees the widening gulf at its starkest. The brightly lit shop-windows are filled with high-tech electronic toys for the children of the rich, and a few yards away, the dark corners of subway entrances are filled with the dim outlines of derelict human beings that the new technology has left behind. In every large American city, and in many cities all over the world, such contrasts have become a part of everyday life. When I first came to America forty-five years ago, rich and poor people were less estranged and less afraid of one another, the feeling of belonging to a community was stronger, the rich had fewer locks on their doors and the poor had roofs over their heads. Since those days, wealth has accumulated and society has decayed. It is as Haldane said, 'The tendency of applied science is to magnify injustices until they become too intolerable to be borne, and the average man, whom all the prophets and poets could not move, turns at last and extinguishes the evil at its source.'

My scientist friends may justly protest that the calamities of American society are caused by drugs, or by guns, or by racial intolerance, or by illiteracy, or by bad schools, or by broken families, rather than by science. It is true that the immediate causes of social disintegration are moral and economic rather than technical. But science must bear a larger share of responsibility for these evils than the majority of scientists are willing to admit. When we look at historical processes on a time-scale of fifty or a hundred years, science is the most powerful driving force of change. Because of science, machines have displaced unskilled manual workers, and computers have displaced unskilled clerical workers, in all branches of industry and commerce. Because of science, the traditionally conservative middle class of well-paid blue-collar industrial workers has almost ceased to exist. Because of science, jobs paying enough to support a family in comfort are no longer available to young people without higher education, unless they happen to be gifted with special talent as baseball players or rock stars. Because of science, families with access to computers and to higher education are rapidly becoming a hereditary caste, the children inheriting these advantages from their parents. Because of science, children deprived of legitimate opportunities to earn a living have strong economic incentives to join gangs and become criminals. In Trenton, only a few miles away from the academic oasis of Princeton where I live, many of the children of the inner city seize their first chance of financial independence at the age of nine, when they are recruited by the drug-dealers to serve as scouts to give warning of police-raids. The big turning-point in their lives comes in the summer between third and fourth grades, when the gang displaces the school as their main source of education. Technological change, driven by science, has been the primary cause of these revolutions in the economic base of society. After technological change has closed down industries and destroyed jobs, the decline of morality and the erosion of discipline follow as secondary causes of social breakdown.

Haldane did not foresee the computer, the most potent agent of social change during the last fifty years. He expected his Daedalus, destroyer of gods and of men, to be a biologist. Instead, the Daedalus of this century turned out to be John von Neumann, the mathematician who consciously pushed mankind into the era of computers. Von Neumann knew well what he was doing. In 1946, after he had been involved for several years in the design of nuclear weapons, he started a project in Princeton to build the first modern computer. A friend asked him in 1946 whether he was still working on bombs. 'No,' said Von Neumann, 'I am working on computers. Computers are much more important than bombs.' Von Neumann, like Haldane's Daedalus, had dreams that went far beyond the scientific instrument that he was building in Princeton. He spoke and wrote much about automata. His automata were abstract generalizations of a computer. An automaton was a machine that could not merely compute but could carry out actions in the real world as instructed by its program. Von Neumann saw that there was no limit to the scale and complexity of actions that automata could perform. His computer was only a small step toward the realization of his dream of automata guided by artificial intelligence. Beyond the intelligent automaton was another dream, the self-reproducing automaton. Von Neumann proved with mathematical rigour that a self-reproducing automaton was possible, and enunciated the abstract principles that would govern its design. He dreamed that the creation of self-reproducing automata would be a boon to mankind, abolishing hunger and poverty all over the earth, providing us with obedient slaves to satisfy our needs. Self-reproducing automata could build our homes, cook our food, and wait on us at table. But Von Neumann, like Haldane's Daedalus, was destined to turn good into evil. If we were malicious, we might say that the ultimate outcome of Von Neumann's dream of self-reproducing automata would be to make all humans superfluous except for mathematicians like himself. In the end, even the mathematicians, who would initially be needed in order to design the automata, might also turn out to be superfluous.

Two developments that Von Neumann did not foresee were the personal computer and the computer-game software industry. These two side-effects of his activities have grown with explosive speed. Like other rapid technological changes, they have brought with them both good and evil. On the good side, they have given us computers with a human face, computers accessible to ordinary people for profit or for fun. Von Neumann never imagined that computers could be humanized to such an extent that mothers would use them to print birth-announcements and schoolchildren would use them to do home-work. On the evil side, the home-computer industry has widened the gap between rich and poor. The child of computer-owning parents grows up computer literate and is showered with opportunities to enter the world of high-tech education and industry. The child without access to a home computer is left behind. Computer illiteracy is an additional barrier that a poor child has to overcome in order to earn an honest living.

Haldane looked unflinchingly at the evil consequences of science, both in war and in peace, but he was not on that account a pessimist. He did not believe in gloom and doom. The final message of *Daedalus* is not gloomy. Haldane had far too much respect for ordinary people to be a pessimist. He admired the toughness of the ordinary soldiers who fought under his command in France. He loved India, and chose to live there at the end of his life, because he had a natural affinity for people who remained cheerful in spite of hardships. When he was in hospital recovering from an unsuccessful cancer operation, he wrote a cheerful poem with the title, 'Cancer's a funny thing'. His innate optimism shines through the superficial cynicism of *Daedalus*. He considered the consequences of science to be mostly evil and danger-ous but, because science forces us to overcome these evils and dangers, he said, 'science holds in her hands one at least of the keys to the thorny and arduous path of moral progress.' The final message of *Daedalus* is that ordinary people can turn evil into good if they have the necessary courage and moral leadership. Haldane had no doubt that they have the courage, and he intended to give them the moral leadership.

Now, seventy years after *Daedalus* was published, two questions remain to be answered. First, when will Haldane's prediction come to pass, that biotechnology rather than computer technology will be the major force for change in human affairs? Second, what should we do in the meantime, in the world as we find it at the end of the twentieth century, to turn the evil consequences of science into good? I will dis-cuss these two questions in turn and give them brief and inadequate answers.

Haldane was surely right in expecting that the most profound shocks to human society would come from biology. He mentioned two biological shocks in particular, the genetic engineering of microbes that would invade the oceans and replace agriculture as a source of food, and the technology of ectogenesis that would replace motherhood as a source of babies. Neither of these shocks has yet happened. Although the technologies of genetic engineering and ectogenesis are developing rapidly and are already in use to a limited extent, giving us new drugs and an occasional test-tube baby, agriculture and motherhood are still alive and well. It does not now seem likely that agriculture and motherhood will be displaced by biotechnology within the next century. Still, there can be little doubt that genetic engineering and ectogenesis are destined to give us rude shocks in one way or another. Sooner or later, genetic engineering will allow us to create new species of plants and animals according to our whim, and to choose the genetic endowment of our children. Ectogenesis will allow us for the first time to achieve full equality of biological status between men and women. Either one of these innovations will bring more profound changes to human society than the advent of personal computers. Both innovations are likely to sharpen the social conflicts between liberal and conservative, between believer and unbeliever, between rich and poor. Both innovations may be resisted successfully for a while by legal restraints, by religious prohibitions, or by physical violence. And in the long run, both innovations are likely to prevail over the opposing forces, at least in some places and in some segments of society. Nobody can predict how long the run will be before these things happen. Haldane expected them to happen before the end of the twentieth century. He turned out to be wrong. I am probably erring on the side of caution when I say that they will happen before the end of the twenty-first.

The final question raised by Haldane's book is, what scientists should do now in order to make science work for good rather than for evil. Here the word 'good' refers only to the social benefits of science. We are not here concerned with the intellectual joys that science provides to scientists, or with the intrinsic excellence of science itself. The ways in which science may work for good and evil in human society are many and various. As a general rule, to which there are many exceptions, we may say that science works for evil when its effect is to provide toys for the rich, and works for good when its effect is to provide necessities for the poor. The phrase 'toys for the rich' means not only toys in the literal sense but technological conveniences that are available to a minority of people and make it harder for those excluded to take part in the economic and cultural life of the community. The phrase 'necessities for the poor' includes not only food and shelter buy adequate public health services, adequate public transportation, and access to decent education and jobs.

The scientific advances of the nineteenth century and the first half of the twentieth were generally beneficial to society as a whole, spreading wealth to rich and poor alike with some degree of equity. The electric light and the telephone and the refrigerator and the radio and the television and the synthetic fabrics and the antibiotics and the vitamins and the vaccines were social equalizers, making life safer and more comfortable for almost everybody, tending to narrow the gap between rich and poor rather than to widen it. Only in the second half of our century has the balance of advantage shifted. During the last forty years, the strongest efforts in pure science have been concentrated in highly esoteric fields remote from contact with everyday problems. Such efforts are unlikely to do harm, or to do good, either to the rich or to the poor. At the same time, the strongest efforts in applied science have been concentrated upon market-driven projects, that is to say, projects that are expected to lead quickly to products that can be profitably sold. Since the rich can be expected to pay more than the poor for new products, market-driven applied science will usually result in the invention of toys for the rich. The failure of science to produce benefits for the poor in recent decades is due to two factors working in combination, the pure scientists becoming more detached from the mundane needs of humanity, the applied scientists becoming more attached to immediate profitability.

Although pure and applied science may appear to be moving in opposite directions, there is a single underlying cause that has affected them both. The cause is the power of committees in the administration and funding of science. In the case of pure science, the committees are mainly composed of scientific experts performing the rituals of peer review. If a committee of scientific experts selects research projects by majority vote, the inevitable result is that projects in fashionable fields are supported while those in unfashionable fields are not. In recent decades, the fashionable fields have been moving further and further into specialized areas remote from contact with things that we can see and touch. In the case of applied science, the committees are mainly composed of business executives and managers. Such people usually give support to projects leading to products that affluent customers like themselves can buy. Only a cantankerous individual like Henry Ford, with dictatorial power over his business, would dare to create a mass market for automobiles by arbitrarily setting his prices low enough and his wages high enough so that his workers could afford to buy his product. Both in pure science and in applied science, rule by committee has the effect of discouraging unfashionable and bold ventures.

Even if we could organize a rebellion against the rule of committees, returning power over pure and applied science to individual scientists and entrepreneurs, this would not guarantee that science would become benign. The rebellion would only create opportunities for more humane uses of science. It would not ensure that the opportunities would be used. To bring about a real shift of priorities, individual scientists and entrepreneurs must use their powers to promote new technologies that are more friendly than the old to poor people and poor countries. The ethical standards of scientists must change as the scope of the good and evil caused by science has changed. In the long run, as Haldane and Einstein said, ethical progress is the only cure for the damage done by scientific progress.

CHAPTER FIVE

HALDANE BETWEEN DAEDALUS AND ICARUS

YARON EZRAHI

In his *Daedalus, or science and the future*, J.B.S. Haldane (1924), moved by his sense of mission as a scientist and increasingly inspired by the socialist vision of a perfect fusion of science and society, presented a progressive view of the future of civilization. 'Science', he wrote, 'is the free activity of man's divine faculties of reason and imagination. . . . It is man's gradual conquest, first of space and time, then of matter as such, then of his own body and those of other living beings, and finally the subjugation of the dark and evil elements in his own soul.' While Haldane was mindful that 'none of these conquests would be complete,' he insisted that 'all . . . will be progressive.' Writing after one world war and before another, Haldane believed that the pains of war were likely to increase popular support for international institutions as means to check the disruptive forces of nationalism.

In a response to Haldane's essay, Bertrand Russell (1924) published *Icarus, or the future of science* in which he questioned Haldane's 'attractive picture of the future as it may become through the use of scientific discoveries to promote human happiness'. Icarus, having been taught to fly by his father Daedalus, 'was destroyed by rashness,' writes Russell. 'I fear', he concludes, 'that the same fate may overtake the populations whom modern men of science have taught to fly. . . . Science is not a substitute for virtue . . . Technical scientific knowledge does not make men sensible in their aims . . . [and] science has not given man more self-control, more kindliness or more power of discounting their passions.'

At the closing of the twentieth century, Russell's caution about the powers of science to transform culture and politics appears more in tune with the experience of the age than Haldane's early optimism. From this perspective, our century's record of massive attempts to apply scientific and technical knowledge to solve technical problems is much more impressive than the no less comprehensive attempts to solve major social and human problems.

This record does not seem to support Haldane's notion of scientific reason as a force progressively conquering new spheres of life where unwarranted beliefs and deep-rooted prejudices are replaced by certified knowledge and human passions are tamed by its force. If anything, the experience of spectacular scientific advances (especially in molecular biology), coupled with declining expectations about the power of science to rationalize social and political life, gives rise to a sense of paradox. The declining faith in the power of science to resolve basic problems of civilization is related both to the disintegration of the Soviet Union which for a long time embodied for Haldane, as for some of his most prominent peers, the hope of fusing science and socialism, and to the disenchantment with attempts to fuse science and democratic politics as well.

While science no doubt greatly influenced public policies and programmes in the American democracy, the record of the attempts to fuse science and democratic politics fell short of the kind of hopes expressed by American progressive thinkers like John Dewey. Also writing between the two world wars, John Dewey (1934) hoped that 'the operation of cooperative intelligence as displayed in science [is] a working model for the union of freedom and authority which is applicable to political as well as other spheres.' As it turned out, attempts to deploy the 'cooperative intelligence' of science as a model for decisions in the political sphere did not and, we now realize, could not, have succeeded.

While no one can deny, then, the impact of science on modern society and politics in both democratic and authoritarian regimes, the expectation that scientific knowledge would replace politics was based on a profound misunderstanding of the very nature and functions of politics. Since Haldane's *Daedalus* was written in a frame of mind still very much shaped by Enlightenment objectives and conceptions of society and politics, it offers more than seven decades later, a valuable opportunity to reflect on the reasons for the gap between Haldane's and his prominent peers' visions of the place of science in future society, and its actual place in this future which at least in part is our present. The questions I would like to address are the following. Why did scientists like Haldane have such an exaggerated idea of the power of science to transform politics and society? Why were they so much off the mark in anticipating the relations between science and politics towards the end of the twentieth century? What has in fact inhibited the Enlightenment programme of rationalizing culture, politics, and society? While, in the years following the publication of *Daedalus*, Haldane changed some of the views he expressed in this essay, it remains useful to evaluate critically his earlier views. A judgment of his orientations in *Daedalus* from the hindsight of the late twentieth century would be particularly valuable in indicating the changes in the relations between science and politics between 1924 and the end of the century.

In some respects, Bertrand Russell's observation that 'science is not a substitute for virtue' already anticipates part of the answer to the above questions. Perhaps it is even more accurate to describe Russell not so much as anticipating the answer as echoing the most influential eighteenth century critic of the Enlightenment belief that the progress of the arts and the sciences will lead to moral progress as well: Jean-Jacques Rousseau (1964). Rousseau charged that the sciences, far from controlling human passions, feed on these passions and serve them. Among other vices, the sciences encourage ambition, irresponsible and dangerous flights of the imagination, and the pursuit of luxury. Moreover, says Rousseau, like the arts the sciences 'spread garlands of flowers over the iron chains with which men are burdened.' Such criticism could be easily directed to British scientists and socialists of the 1930s like Haldane and Bernal who provided intellectual scientific legitimation for the Soviet regime, focusing on the garlands yet ignoring for too long the iron chains which were hidden behind the flowers (Wersky 1988). Most germane to our subject and to our enquiry into what Haldane took for granted (at least in his *Daedalus*) is Rousseau's concern with the transfer of scientific knowledge to the sociopolitical context. 'Even with the best intentions, by what signs' he asked, 'is one certain to recognize it? In this multitude of different opinions what will be our criterion in order to judge it properly? And hardest of all, if by luck we finally find it who among us will know now to make good use of the truth?' Rousseau is mindful here of the fact that most citizens lack the education and the tools by which scientists can distinguish valid from false ideas and of the even greater difficulty of ensuring that available scientific truths will be put to good use in the context of public affairs. It is precisely because Rousseau (1950), was aware of the inherent constraints on the uses of knowledge to discipline public passions and opinions that he insisted on educating the public to respect the dogmas of what he called 'civil religion' in order to uphold the structures of democratic politics. Rousseau, who was deeply concerned about the power of the human imagination to drive people to pursue unrealizable, and therefore enormously costly, fantasies, thought that science and faith in the power of science to revolutionize life radically might feed such fantasies and threaten the stability of the sociopolitical order. By contrast, advocates of the Enlightenment, from Condorcet to Haldane, have believed in the power of science to control human passions and ground society in rational knowledge and freedom. Haldane clearly thought, at least in 1924, that scientific knowledge is potentially the means with which humanity can achieve the freedoms of self-control and self-regulation. It is, however, precisely because Rousseau, who was at least partly despirited by Montaigne's contagious scepticism, did not trust the restraining effects of reason, that he preferred the simplicity and austerity of Sparta as a model for the good society over the culturally enterprising Athena, and elevated the humble peasant who lives close to nature to the intellectual as a model of the virtuous citizen.

The principal flaw in Haldane's elitist humanist scientism seems to lie in his conception of values and their relations to politics. While Haldane's essay lends itself in some key places to various and even contradictory readings and interpretations, it seems to me that his faith that science's effects on society and politics, with important exceptions, are essentially progressive, is rooted in his belief that ultimately people can agree on conceptions of the good and thus generate public mandates for the application of science to the promotion of their supposedly shared goals. In such a universe, politics and even the state could be dispensable, as Marxism expected. But if politics in a free society is in essence at least partly the process through which the diverse, and often contradictory or incommensurable, interests and preferences of the citizens are articulated, and areas of agreement and disagreement among them are delineated for the purpose of, temporarily, fixing the parameters of legitimate public policies and actions, then the only alternative to politics is coerced agreement. The history of the interaction between science and politics in twentieth century western societies seems to indicate that the unresolvable nature of basic conflicts of values and interests, the absence of a consensual, compelling basis for choosing among competing value orders, constitutes a formidable constraint on the fullest use of scientific and technical knowledge in public affairs.

Under such conditions, politically negotiated coalitions and compromises are the only methods of collective decision-making which are consistent with democratic principles of order and freedom. But inevitably such coalitions and compromises generate inherently ambiguous, and often self-contradictory, mandates of action which are not congenial for the smooth application of science and technology to the promotion of public goals. It is easy to appreciate the impulse of scientists who, like Haldane, have visions of what scientific knowledge could accomplish, to search for ways to transcend the fuzzy compromises of democratic politics and dream of a normatively homogeneous society in which no such differences and conflicts of values and interests hamper the full power of science to do good. Haldane held that the future would be better 'if mankind can adjust its morality to its powers. If we succeed in this then science holds in her hands one, at least, of the keys to the thorny and ardorous path of moral progress. . .'. Here Haldane reveals a hidden premise characteristic of central ideologues of science since the Enlightenment: that value conflicts are largely the result of ignorance and can somehow be resolved by the common knowledge of scientific truths. Science is seen not only as a potentially non-controversial means to realize shared goals but also as the very basis for the formation of values and ethical consensus. While we are referring here to views expressed by Haldane in the 1920s, such attitudes persisted in large parts of the scientific community many years later. Almost half a century after the publication of *Daedalus*, the French biologist Jacques Monod (1971), for example, used the expression 'ethics of knowledge' to indicate the scientists' responsibility to seek the substitution of knowledge for conventional beliefs and help to 'lay the foundations for a value system wholly compatible with science itself and able to serve humanity in its scientific age'. Implicit in this view is a belief that scientific rationality can and should be a model for rationality in human affairs. The 'ethics of knowledge' defines it as the duty of the scientist-qua-scientist to advocate and promote the role of science as the basis for the social perceptions and manipulations of 'reality'. It deduces from the scientist's internal-professional commitments to the canons of science a generalizable commitment to the realization of a vision of society based on scientific rationality.

The extension of the intra-professional ethics of knowledge to the entire society is of course unacceptable and impractical in a free society in which the spontaneous expression of values and preferences is unlikely to produce a consensus or decisively privileging the value of knowledge, or for that matter any other value, over all other values. The modern democratic experience indicates that the free expression of ideas, interests, and preferences does not normally result in producing uniform or unified but rather a wide diversity of opinions and goals. It is common in modern democracies that government decision-makers receive from their constituencies mandates to take two or more alternative and often conflicting courses of action serving equally cherished and supportable objectives. In such contexts it is not at all clear what the meaning of such statements as putting 'science to work in the service of humanity' would be.

Against this background, it is not difficult to see how scientists so deeply committed to the scientific ethos of the Enlightenment as Haldane, Bernal, and Needham could dream of a normatively homogeneous society in which a presumably universal agreement on a conception of justice and basic social goals allows a massive application of science in all fields of human action, and why they could be attracted to socialism and to the idea that a communist system like the Soviet Union was developing into such a society. Marxism evolved a vision of a community liberated from social conflict which could dispense with the state as a necessary means for the resolution of such conflicts and leave room for the guiding role of knowledge. It appeared compatible with the idea, so popular within the scientific communities of the period, that public politics can and should be depoliticized as science replaces politics in directing collective choices and actions.

Like that of many of his peers, Haldane's readiness to assume the possibility of a homogeneous value environment for an unproblematic application of science to social problems, the belief that criteria of scientific and technical rationality could be universally applicable outside the context of science, may have been encouraged by the fact that science has historically emerged as an enterprise controlled by a shared professional commitment to the priority of the advancement of systematic knowledge about the world over all other goals or values, a commitment which generates within the scientific community a relatively high consensus on goals, means, and criteria for judging or evaluating professional activities such as research. Insofar as the advancement of knowledge takes precedence over such other goals as profit-making, constructing more efficient or effective machines, solving pressing practical individual and social problems, etc., the normative orientation of pure science or of the basic research community can be distinct and relatively homogeneous in relation to the more heterogeneous and wide-ranging orientations of the larger society. The political community, in a free society, is organized, by contrast, around a much more discrete and wider set of often internally inconsistent goals, none of which can remain fixed for a long time. Modern attempts to fix ideologically the value structure of a whole society, exemplified in the former Soviet Union, are fundamentally at odds with the political process in a democratic society which is continually engaged in never-ending attempts to order and re-order goals and interests. Such goals as the advancement of science, to be sure, have been recognized by most democracies as important although never as the highest and most important goal. Scientists have been often oblivious to the fact that not only when the status of science in society is low but also when it is high, it is ideology, which informs the attitudes of the lay public towards science, that decides the matter. While the scientific community differentiates itself from the larger society, including the private and the public sectors, so that it can maximize professional scientific goals without being strictly and directly subject to external market or political tests, the polity as the most inclusive social association, must constantly balance such goals as the advancement of knowledge, the maintenance of social services, and the maintenance of order. While such goals can at times partially converge, they are more often competitive for scarce material and political resources. In such a context, the expectation that goals can be fixed in a clear order cannot be satisfied without violating basic rules of democratic politics. This is partly why 'Five Year Plans' of the kind generated in the former Soviet Union are not a likely government technique in democracies where shifting compromises in an open political competition render long-range policy planning very difficult.

Given the differences between the pluralistic competitive value environment of democratic public policy-making and the relatively homogeneous consensual value environment of the professional activities of science, any attempt to deploy scientific standards of reliability, validity, and rationality in the larger context of public affairs is likely to be ridden with endless difficulties and frustrations. It has been very difficult for scientists to recognize that concepts and theories of nature and society which are selected and sustained as scientifically certified because of their relatively advantageous explanatory power, predictive value, etc. cannot be unquestionably accepted in the political context of the larger society on the same basis. That is, in the larger society concepts and theories of nature and society are bound to be evaluated not just with reference to strict intellectual criteria, but also with reference to other kinds of considerations such as their anticipated or perceived impact on socially prevailing religious or ideological conceptions of nature and society as well as their influence on established cultural institutions, educational practices, etc. The ongoing difficulties in the reception of Darwinian theories in biology and the controversy around the linking of IQ scores and genetics in the United States during the 1970s are illustrative. These controversies demonstrate the interactive impact of scientific ideas in the sociopolitical context which scientists have often regarded as illegitimate.

Haldane was aware of the fact that the effects of relentless scientific reason can shake important, even sacred, social symbols, conventions, and values. But he welcomed that process. 'Consider so simple and time honoured a process as the milking of a cow,' he observed in *Daedalus*. 'The milk which should have been an intimate and almost sacramental bond between mother and child is elicited by the deft fingers of a milkmaid, and drunk, cooked, or even allowed to rot into cheese. We have only to imagine ourselves as drinking any of its other secretions, in order to realize the radical tendency of our relation to the cow.' Haldane believed that somehow an inevitable process of scientific progress would eventually force ethical systems hostile to such scientific innovations to adjust. Referring to the sacred cows of India, he noted that 'the holiness of the cow has unfortunately extended to all its products . . . it would perhaps be better,' he concluded, 'if milking could be secularized'. The Haldane who wrote *Daedalus* clearly believed that it is better to treat milking in terms of instrumental rationality than religion and that scientific knowledge and techniques are inherently more progressive *vis-à-vis* existing social conventions, religious practices, etc. This attitude clearly ignores the fact that 'science' and 'secular attitudes' represent here alternative ethical or value systems and that the choice between them cannot be regarded as unproblematically warranted by the judgment that it is more 'progressive', a judgment made in terms of one alternative value system itself. Haldane, moreover, ignores the political logic of ordering and re-ordering values and the related competition among alternative theories, conceptions, and practices in the larger social context. Thomas Hobbes (1962), who was both inspired by early science and critical of leading scientists of his day like Robert Boyle, wrote in a revealing section in his *Leviathan* that scientific concepts of nature must at times be rejected as a basis of public affairs when they appear to undermine social and political attitudes that are essential for upholding peace and order. 'If nature . . . has made men equal,' he wrote, 'that equality is to be acknowledged; or if nature made men unequal; yet because men that think themselves equal will not enter into conditions of peace but upon equal terms such equality *must be admitted*' (my emphasis).

Hobbes understood something which Haldane and his circle ignored. The political order is partly upheld by important regulatory fictions, commitments, and conventions whose erosion can be very costly to both stability and freedom. In modern democracies the idea of basic equality in the natural endowment of diverse individuals and groups and the idea that individuals are autonomous agencies capable of voluntary actions have had a constitutive function in the political and legal systems. Occasional attempts to question these ideas by reference to various biological, psychological, and behavioural theories have, therefore, characteristically provoked resistance in both the political and the legal systems. Such reactions to attempts to modify educational policies in the light of data on differences in the average IQ scores of various ethnic groups, or to draw from studies in human behaviour or sociobiological research conclusions which question the idea of autonomous agency, illustrate the potential tensions pointed out by Hobbes between scientific assertions and 'conditions of peace'. This applies, of course, also to such phenomena as Haldane's eugenic ideas about selective breeding and his concerns with the 'fertility of the less desirable'.

Applying narrow scientific standards of acceptability to ideas, concepts, or practices in the political context of public affairs often involves, therefore, committing what I have called elsewhere 'the fallacy of misplaced rationality' (Ezrahi 1976). It is characteristic of scientists who commit this fallacy to interpret negative reactions to the application of strict scientific standards to ideas which serve explicit or latent social and political functions, as resulting from ignorance, irrationalism, or prejudice. Sometimes, of course, scientific criticism of conventions or ideologies which help current social solidarity and justify public authority appear appropriate when they are directed against the use of scientific authority to back up racist policies like those in Nazi Germany, or the disastrous policies following the rise of Lysenko in the USSR. But people approve of such scientific criticisms of politically established ideas and practices in the larger sociopolitical context not just because the latter are scientifically false but because they appear to be morally and politically unacceptable to them. Because moral and political criticisms can be, and usually are, independent of strictly scientific criticism, scientifically valid ideas and practices can be morally and politically unacceptable as grounds of public decisions and actions, and scientifically unwarranted ideas can be approved on moral and political grounds as some concepts of equality in western democracies may illustrate. When Haldane failed to appreciate the possibility that the application of advanced scientific knowledge could be morally and politically unacceptable, he tended to think of the application of scientific knowledge as almost inevitable and suggested that moral values should be flexible enough to adapt to the new conditions. He insisted, for instance, that 'a religion [one may also add an ideology], which will frankly admit that its mythology and morals are provisional . . . is the only sort of religion that would satisfy the scientific mind. . .'. Such faith in the task of 'relentless scientific reason' to conquer all spheres of life, ignores, of course, not only natural spiritual needs but also the role of ideology and politics in such important matters as justifying public authority, agreeing on standards of fairness in relation to competing interests and groups, or respecting prevailing emotions and moral sentiments. The idea that scientifically valid theories and practices should be applied simply by virtue of being scientific, in fact licenses the privileging of some values, which such applications selectively serve, at the top of an order of values and thus forecloses the possibility of a free, open-ended contest among diverse normative perspectives on public allegiance.

Given the heterogeneous normative context of politics in an open pluralistic democracy, any body of knowledge has a political dimension which consists of its potential for actual symbolic as well as material effects on the relative positions of competing groups and interests. Hence the same body of knowledge may have different political imports in different contexts. Moreover, the contextual effects or uses of any body of knowledge are not likely to be politically neutral in the sense of being symmetrical with respect to the existing positions of the contending actors and interests. This 'political relativity' of the interactions of scientific knowledge in the context of public affairs does not necessarily challenge the professional intellectual integrity of the men of knowledge, nor does it entail a cognitive epistemological or logical relativism or call into question the internal validity of the involved scientific theories. Again we are talking about relations and configurations formed between fragments of distinct universes of discourse and action.

Since Haldane wrote *Daedalus*, at least western scientists have learned to refrain from automatically classifying any objection to the application of scientific knowledge in human affairs as an 'abuse of science' or an expression of 'political irrationality'. It has become increasingly clear that science and politics serve at least initially very different purposes, that science cannot serve some of the key and most vital goals in relation to which politics is instrumental, that each has a relatively autonomous set of norms and institutions, and that what is acceptable in the context of the one does not necessarily have to be compatible with fundamental ideas and concepts in the other. It has become, therefore, increasingly appreciated that science and politics can cooperate, following careful and elaborate negotiations which factor in the respective differences in their orientations, standards, vocabularies, and methods of action, and that it is necessary to recognize that the external-contextual political profile of each body of knowledge or skills is no less relevant to its prospect in the context of public affairs than how many, or how strongly, scientists support that body of scientific knowledge and skills as the most relevant or effective in dealing with a given problem.

Still the danger of either science or politics trying to reduce each other to their own terms is aggravated by the ambiguities concerning the demarcation lines between them. Robert M. Young (1971) must have been mindful of this problem when he observed that 'there is not at any point any clear line of demarcation between pure science, generalizations based on it, and the related theological, social, political and ideological issues.' While it is difficult to keep science and politics sufficiently distinct under such conditions so that each can combine with the other without losing their integrity, such conditions facilitate the urge, so powerfully manifest in Haldane, to find or construct an all-encompassing scientific world view. Haldane argued again and again that science must march forward forcing the imaginary universes of religion, philosophy, metaphysics, and politics to adapt to the compelling force of scientific knowledge of the 'real' universe. Instead of respecting the differences among domains of culture, society, and politics, he argued in 1924 for a total unified world in which, among other things, contemporary poets, like their earlier counterparts, Milton, Shelley, and Keats, would be educated in science.

During the second half of the twentieth century, criticism of notions of culture based on such culturally reductionist scientific realism have been reinforced by increasing appreciation for the role of social factors in the production and deployment of conceptions of reality including those which are attributed to science (Durkheim 1965). It is better understood now that there is no such thing as scientific or technological influence, on culture, society, or politics which are not mediated explicitly or implicitly by ideology and, therefore, there is no escape from the need to make extra-scientific judgments before science can have, or be deprived of having, such influence. Robert M. Young (1971), speaking under the sponsorship of the British Society for the Social Responsibility of Science in 1970, observed astutely that scientists need for their moral purposes 'to think seriously about the metaphysics of science, about the philosophy of nature, of man and of society, and especially about the ideological assumptions which underlie, constrain and are fed by science. Since we have systematically weeded out this tradition among working scientists—one which flourished until the 1920s—we need help from other disciplines in gaining the necessary perspective and we could well turn to the continuing traditions of enquiry in the social and political sciences.'

Since scientific concepts, ideas, techniques, and authorities are inevitably converted into political resources once they are deployed in the competitive sphere of public affairs, it is necessary, as Hobbes recommended, to evaluate the ethical and political effects of scientific approaches to problems which arise in the social and the political context. While within the boundaries of scientific institutions free enquiry should not be restricted, the deployment of the results of scientific enquiry and their applications in the context of public affairs cannot skip the process of ethical, political, and social selection. To acknowledge this is in fact to refuse to dream with Haldane of 1924 that 'in the future perhaps it may be possible by selective breeding to change character as quickly as institutions.' He seemed to have felt then that it was just a matter of overcoming the resistance of the ignorant or the uninitiated. Such a belief that scientific reason can constitute the foundation of a modern progressive society has not survived into the end of the twentieth century. It is precisely because modern democracies accommodate diverse types of individuals, groups, cultures, and institutions and 'man's capacity,' as Haldane put it, 'to subjugate the dark and evil elements in his own soul' is not so widely trusted, that these democracies have evolved complex legal and institutional mechanisms for checking excessive uses of public and private powers. From the fact that human beings have learned to control the energies of the atomic nucleus and can in principle try to direct their own evolution, it does not follow that mankind has solved or is likely to solve in the future the problems of self-restraint, nor that there is any reason to hope for, or even desire, a universal agreement or a common system of values. Since people like Haldane believe that human biology can and will be changed by deliberate applications of scientific knowledge, how can any idea of a fixed given human nature, in the biological sense, even theoretically ground such a universal ethics? At the end of the twentieth century it is becoming increasingly clear that there is neither scientific knowledge nor ethics nor values which can provide a sufficiently stable and unambiguous basis for a massive application of biology that can self-evidently appear as 'serving humanity'. The process of applying biological knowledge in medicine, agriculture, and other areas is, to be sure, often revolutionary and matches some of Haldane's most daring predictions. But it is clearly inhibited by a deeper appreciation for the unanticipated results of biological interventions, including their interactions with human behaviour in sensitive areas, and the need to respect the values and ethical commitments of diverse religious and social groups which Haldane would have considered primitive or regressive.

It appears that while in the period since Haldane wrote *Daedalus* the advancement of science and technology has been spectacular, science has lost much of its earlier lustre as a symbol of, and a central element in, 'high culture'. This change can be traced in part to the discrediting of the very idea of 'high culture' in modern democracies. But no less importantly, it reflects a growing respect for the inherent diversity of human values and for the fact that the Enlightenment 'ethics of knowledge' is not regarded as compellingly privileged relative to other ethical orientations nor that men of knowledge have unchallengeable advantage in choosing between alternative value systems. Nowadays science is part and parcel of a democratic culture with powerful links to material and popular cultures as well as, of course, to the universities and the government. Looking from the elevated perspective of science as high culture in the early part of the century, the place of science in late twentieth century culture and society would have probably discouraged Haldane. Haldane was concerned (1927), that 'as the ideals of pure science become more and more remote from those of the general public, science will tend to degenerate more and more into medical and engineering technology, just as art may degenerate into illustration and religion into ritual when they lose their spark. . .'.

To lose the spark of pure science was to Haldane also to lose its unique cultural authority. He held that the future of western civilization depends upon whether or not it can assimilate that scientific point of view. But from the perspective of the end of the twentieth century, one does not need to discard the importance of the scientific outlook in order to recognize that the process of democratization has put it on par with other perspectives which must be considered. Today's reader is not likely to be as inspired by Haldane's romantic vision of the mission of the solitary scientist to advance the conquest of 'relentless reason' in the wider spheres of culture and society. But Haldane's point about the special responsibility of the scientist, if not to apply scientific knowledge against all odds at least to inform, to warn, and to enlighten his audience without yielding to economic and political pressures, is as valid today as it was when he wrote *Daedalus*.

REFERENCES

Dewey, J. (1934). Science and the future of society. In *Intelligence in the modern world* (ed. Joseph Ratner), p. 360. Modern Library, New York.

Durkheim, E. (1965). *The elementary forms of religious life* (trans. J.W. Swain). The Free Press, New York.

Ezrahi, Y. (1976). The Jensen controversy: A study in the ethics and politics of knowledge in democracy. In *Controversies and decisions* (ed. Charles Frankel), pp. 149–50. Russell Sage Foundation, New York.

Ezrahi, Y. (1990). *The descent of Icarus, science and the transformation of contemporary democracy*. Harvard University Press, Cambridge, MA.

Haldane, J.B.S. (1924). *Daedalus, or science and the future*. E.P. Dutton and Co., New York.

Haldane, J.B.S. (1927). *Possible worlds and other essays*. Chatto & Windus, London.

Haldane, J.B.S. (1932). *The inequality of man and other essays*. Chatto & Windus, London.

Hobbes, T. (1962). *Leviathan* (ed. Michael Oakeshott), p. 120. Blackwell, Oxford.

Monod, J. (1971). On the logical relationship between knowledge and values. In *The Social impact of modern biology* (ed. Watson Fuller). Routledge and Kegan Paul, London.

Rousseau, J.J. (1964). *Discourse on the question: has the restoration of the sciences and the arts tended to purify morals?* (1750). In *The first and second discourses* (ed. R.O. Masters). St. Martin's Press, New York.

Rousseau, J.J. (1950). *The social contract*. In *The social contract and discourses* (ed. and trans. G.D.H. Cole). E.P. Dutton and Co., New York.

Russell, B. (1924). *Icarus, or the future of science*, pp. 5, 6, 55, 62, 63. E.P. Dutton and Co., New York.

Young, Robert M. (1971). *Evolutionary biology and ideology: then and now*. In *The social impact of modern biology* (ed. Watson Fuller), pp. 205, 211. Routledge and Kegan Paul, London.

Wersky, G. (1988). *The invisible college, A collective biography of British scientists and socialists of the 1930's*. Free Association Books, London.

CHAPTER SIX

HALDANE'S *DAEDALUS*

ERNST MAYR

In 1923, the 31 year old polymath, J.B.S. Haldane published an essay (93 pages) with the revealing subtitle 'Science and the future' that was the sensation of the day. It had six printings the first year, another four in the next two years, and had an immense impact owing to its wide distribution. One of its ideas, the artificial production of babies, became the major theme in Aldous Huxley's book *Brave new world*.

In spite of its striking originality, *Daedalus* was a product of its time. Prophesies about the future had become popular ever since H.G. Wells' writings. The First World War seemed to have destroyed many of the Victorian constraints and ushered in a new spirit of liberation and experimentation. It encouraged bold minds to test new directions and to question what everybody had taken for granted up to then.

Inspired by Wells, Haldane indeed makes numerous predictions for the future and *Daedalus* is often considered simply a contribution to the futurology literature. However, to look at *Daedalus* only from this point of view would be a great mistake. For Haldane, *Daedalus* was not simply a piece of literature, but rather a manifesto of some of his deepest beliefs and hopes. Two aspects of this work are particularly impressive: Haldane's optimism for a better future for mankind, and his conviction that science, particularly the science of biology, would lead us to this desired goal.

Reading Haldane is quite inspiring owing to his optimism, but it is also lots of fun. He never misses an opportunity to make fun of idols and to kill sacred cows. And he loves to make comparisons that are bound to flout other people's sensitivities, as when referring to Einstein as 'the greatest Jew since Jesus' (p. 26).

In recent years we have experienced a rather widespread hostility against science and have assumed that this is a modern development.

It comes rather as a surprise to find that Haldane himself felt called upon in 1923 to defend science against numerous attacks. But ultimately, says Haldane, 'scientific research has little to fear' because both capitalism and labour depend on the products of research. There is no danger that science will reach an impasse, and 'to show how far from complete is any branch of science at the present time' is one of Haldane's objectives. For Haldane, as it was for Vannevar Bush many years later, science was 'an endless frontier'.

But what will these future developments of science be? Haldane touched the subject of physics and its application only briefly. However, he says it is safe to say that a better knowledge of the properties of radiation will permit us to produce light in a more satisfactory manner than is possible at present. By discovering new methods of producing light by other means than hot bodies (which is very wasteful), it will be possible to produce light much more cheaply, and he predicts 'in fifty years light will cost one fiftieth of its present price.' Even though we now have fluorescent light, the price of light has come down only very slowly and not through the discovery of any drastically new source of light. However, Haldane also predicts great improvements in transportation (travel) and communication and here his basic predictions certainly have come true. Greatly speeded up rates of travel would shrink the world and would 'bring mankind more and more together,' to make the world one world. Such a development would be 'rich in possibilities—to increase indefinitely man's power for good and evil' (p. 29). It is evident that Haldane was fully cognizant of the dangers of the dream of one world, as subsequently documented by the spread of nuclear weapons and of diseases like AIDS.

All industry depends on energy and Haldane realized, as did everyone else, that oil and coal supplies would be exhausted within centuries. However, says Haldane, this will not 'lead to the collapse of industrial civilization'. Instead, 'we shall have to tap those intermittent but inexhaustible sources of power, the wind and the sunlight. The problem is simply one of storing their energy' (p. 30) and Haldane works out in considerable detail a plan to produce energy with windmills and store the temporary excess energy as liquefied gases. One advantage of such a system would be that air pollution would be prevented. Haldane even gives consideration to radioactivity but does 'not much believe in the commercial possibility of induced radioactivity' (p. 31). This was, of course, years before one knew about nuclear fission and fusion.

Applied chemistry, says Haldane, has not up to now 'introduced into human life [any] radical novelty of the importance of the heat engine or the telegraph' (p. 32), but Haldane predicts a great future for this field. There are two classes of desirable chemical substances. One class, such as iron, wood, or glass, is useful on account of its physical or chemical properties. The other kinds of substance 'are desirable on account of their physiological properties. Such substances include food, drinks, tobacco, and drugs.'

Stimulants seem to have intrigued Haldane a good deal, particularly caffeine and nicotine, perhaps an obsession in his social circle, as indicated by the later experiments of his friend Aldous Huxley with various hallucinogenic drugs. Haldane hoped, apparently, that eventually 'mankind will ultimately have at his disposal a vast array of substances like wine, coffee, and tobacco, whose intelligent use can add to the amenity of life and promote the expression of man's higher faculties' (p. 34). No thought is given to addiction and the probable abuse of such substances, which have had such a devastating impact on modern civilization.

Haldane correctly prophesied the shift from an agricultural to an almost completely industrial society that has taken place during the past 70 years. But he saw the new society as a stable society, not envisioning the periods of boom and depression that actually took place. And the people that produced this industrial society 'will inherit the earth. In sum, I believe that the progress of science will ultimately make industrial injustice as self destructive as it is now making international injustice' (p. 29).

'The consequences will be that agriculture will be a luxury and that mankind will be completely urbanized.' Curiously, Haldane was remarkably prescient about this sociological development, but not about its causation. When it occurred it was not because food production was taken over by factories, but rather by agriculture itself having become mechanized.

Haldane's sympathies were clearly with the city, with industry, and with factories. 'Personally, I do not regret the probable disappearance of the agricultural labor in favour of the factory worker, who seems to me a higher type of person from most points of view' (p. 34).

Scattered through these discussions are expressions of Haldane's socioeconomic and political views, and this led him to make economic predictions. So he considers it 'quite possible that capitalism itself may demand that the control of certain key industries be handed over completely to the workers of these industries simply in order to reduce the number of sporadic strikes in them' (p. 29). Evidently, Haldane took it for granted that the workers in their infinite wisdom would do a fine job. He did not live to see the collapse of the worker controlled industries in the soviet countries.

Although physics was the dominant science in the past, Haldane is quite convinced that by now 'the centre of scientific interest lies in biology' (p. 26). This was claimed two years before the pronouncement of the theory of quantum mechanics. He was quite confident about the impact of science on beliefs and policies: 'Darwin's results are beginning to be appreciated, with alarming effects on certain types of religions, those of Weismann and Mendel . . . are going to affect political and philosophical theories almost equally profoundly. I need hardly say that these latter results deal with the question of reproduction and heredity' (p. 38). I am afraid that Haldane was far too optimistic. The findings of genetics and demography have had remarkably little impact on religions, philosophies, or politics. The actions of world governments *vis-à-vis* the population explosion is sad evidence for this. Haldane himself may have expected this, for he ends his discussion with a somewhat defensive statement: 'But owing to man's fortunate capacity for thinking in watertight (or rather idea-tight) compartments, [those shocks to our beliefs] will probably not have immediate and disruptive effects upon society any more than Darwinism had' (p. 38).

Haldane's love for biology was a rather late development and not at all reflected in his education. In college he had concentrated on the classics and a little chemistry, and was officially a biochemist until the early 1930s. Yet, with his brilliant mind he was able to acquire, through attending courses and a great deal of reading, a broad knowledge of biology and an understanding quite beyond that of many of the actual practitioners of the field. Nevertheless, he hardly had the authority to be able to claim that 'we are at present almost completely ignorant of biology . . . actually biological theory consists of some ancient but not very easily stated truths, about organisms in general, due largely to Aristotle, Hippocrates, and Harvey, a few great principles such as those formulated by Darwin, [Robert] Mayer, Claude Bernard, and Mendel, and a vast mass of facts about individual organisms and their parts which are still awaiting adequate generalization' (p. 38). To be sure, the evolutionary synthesis and the rise of molecular biology were still in the future, but there were vast branches of biology, particularly physiology, comparative anatomy, embryology, cytology, and systematics that clearly refute his claim of an 'almost complete ignorance of biology'. And yet, Haldane had enormous faith in biology: 'A time will come when physiology will invade and destroy mathematical physics' (p. 27).

The conceptual advances of biology, thought Haldane, had been far eclipsed by the 'practical applications of biology'. 'I believe that the progress of medicine has had almost, if not quite, as deep an effect on society in western Europe as the industrial revolution. . . . Whereas 400 years ago most people died in childhood, they now live on average . . . until 45.' As a result, 'death has receded far into the background of our normal thoughts' (p. 39). This was stated 15 years before the antibiotic revolution that has made such a change as far as death from bacterial diseases is concerned. And, largely owing to this advance, life expectancy in civilized countries has now risen from 45 years to over 70 years.

Haldane did not prophesy some of the major events in recent medical history, for example the total eradication of small pox, one of the greatest scourges of mankind, and the possibility that an entirely new devastating disease could develop, such as AIDS. With his optimism, if such a possibility had been suggested to him by someone he would almost certainly have predicted that medical science would defeat this new disease in short order.

After a short reference to inter-planetary communication which, says Haldane, will surely be attempted, he finally takes up the application of biology to human life. Haldane is rather scornful of the 'average prophet' whose predictions are far too tame for Haldane's taste. They include 'possibly the introduction of a little eugenics'. He viciously pokes fun at such endeavours. 'The eugenic official, a compound, it would appear, of the policeman, the priest and the procurer, is to hale us off at suitable intervals to the local temple of Venus Genetrix with a partner chosen, one gathers, by something of the nature of a glorified medical board.' Such a procedure, says Haldane, reveals 'a lack of originality and knowledge of human nature'. Instead, says Haldane, 'the ends proposed by the eugenicist will be attained in a very different manner.'

Before going on with the report of Haldane's ideas, it is necessary to say a few general words about eugenics. When one reads the literature of the first decades of this century, one is amazed at the virtually universal popularity of eugenics. It was supported by writers from the far left, all across to those of the far right. Enough was understood about the nature of inheritance to support the realization that a genetic improvement of mankind would be possible only if the best endowed human beings had the greatest reproductive success. And, of course, this is still true even after the demise of eugenics.

But back to Haldane. He was scornful of the humdrum proposals of other eugenicists, and was apparently himself afraid of proposing a shockingly novel scheme of approach. He therefore uses the ploy of an essay of a college undergraduate 150 years from now, reporting on the development of eugenic practices during the twentieth century. The great development was that of what is now called a test tube baby. 'It was in 1951,' reports the essay, 'that Dupont and Schwartz produced the first ectogenetic child.' An ectogenetic child is a fertilized human egg kept alive and growing in appropriate solutions 'for 9 months, and then brought out into the air'.

Curiously, in 1923, the possibility of a dangerous overpopulation had not yet been recognized and (for Haldane) ectogenesis was the answer for the falling birth rate, 'for the birthrate was already less than the death rate in most civilized countries.' Not surprisingly, in his utopia Haldane reports that 'France was the first country to adopt ectogenesis officially and by 1968 was producing 60,000 children annually by this method. As we know, ectogenesis is now universal, and in this country less than 30% of children are now born of women.'

At the time, Haldane's proposal caused a tremendous outcry and scandalized just about everybody. Now, just two generations later, the artificial fertilization of human eggs has become a routine matter except that they are subsequently implanted into their own or surrogate mothers rather than kept in a nutrient solution for the entire period of nine months. However, Haldane's myth inspired his friend Aldous Huxley to elaborate the scenario in his dystopia *Brave new world* (1932).

Haldane thought that his ectogenetic scheme was an excellent way to carry out eugenics. 'The small proportion of men and women who are selected as ancestors for the next generation are so undoubtedly superior to the average that the advance in each generation . . . is very startling. Had it not been for ectogenesis there can be little doubt that civilization would have collapsed within a measurable time owing to the greater fertility of the less desirable members of the population in almost all countries.'

Haldane nowhere mentions in this account on the basis of what criteria the members of the commission would be appointed who do the selecting, nor how the superiority of the germ donors would be established. He apparently had no awareness of the practical difficulties and is quite uninhibited in his predictions. 'If reproduction is once completely separated from sexual love, mankind will be free in an altogether new sense,' or, 'in the future perhaps it may be possible by selective breeding to change character as quickly as institutions,' or, 'we can already alter animal species to an enormous extent, and it seems only a question of time before we shall be able to apply the same principle to our own.'

In his *Causes of evolution* (1932), Haldane was already far less enthusiastic about eugenics and, of course, Hitler sounded the death knell of all eugenic thinking. It is an irony of history that the first country supposedly to adopt eugenics, Nazi Germany, adopted instead something that was not at all what the proponents of eugenics had proposed. The Nazis were typologists, as were so many German biologists at that time, and were simply carrying out typological genocide, not a selective breeding of individuals based on superior qualities. This has been well recognized in a recent analysis of the history of eugenics (Adams 1990).

There is little doubt that man reached humanity through natural selection. And it is equally true that as yet we know of no other method than selection to improve the human genotype. We have methods of eliminating deleterious genes and genetic engineering may be able to perfect methods to introduce certain desirable genes to the genotype, but superior achievement is the result of a fortunate interaction of so many different genes that no artificial method has yet been developed that could achieve what, it was believed, eugenics would accomplish. I have called attention to our difficulties on a previous occasion: 'To apply artificial selection to man is impossible, at least for the time being, for a number of reasons. The first one is that it is quite unknown to what extent nonphysical human characteristics have a genetic basis. Second, human society thrives on the diversity of talents and capabilities of its members; even if we had the ability to manage the selection, we would not have any idea for what particular mixture of talents we should strive. Finally, the concept that people are genetically different, even were it scientifically even better established than it is today, is not acceptable to the majority of western people. There is a complete ideological clash between the concepts of egalitarianism and eugenics' (Mayr 1982, p. 624).

Haldane realized that the genetic approach is not the only one. 'There are perhaps equally great possibilities in the way of the direct improvement of the individual, as we come to know more of the physiological obstacles to the development of different faculties. But at present we can only guess at the nature of these obstacles and the line of attack suggested in the myth is the one which seems most obvious to a Darwinian.' Haldane then discusses treatment with hormones in order to modify psychological conditions and deals more generally with possibility of progress in medicine. Here he failed to predict the revolution caused by the introduction of antibiotics. Yet, he realized that cancer was a disease quite different from the infectious diseases. He was very optimistic about the future of medicine and thought that in due time, medicine would be able to cope with all diseases. Then, 'the abolition of disease will make death a physiological event like sleep.'

In his concluding discussions, Haldane states that he would be 'amply repaid if he had convinced any one person that science has still a good deal up her sleeve'. He comes back to his earlier claim of the importance of the work of the biologist. 'I believe that the biologist is the most romantic figure on earth at the present day.' He refers to the great intellectual revolutionaries of the past like Voltaire, Bentham, and Marx, and adds, 'that Darwin furnishes an example of the same relentlessness of reason in the field of science. I suspect that as it becomes clear that at present reason not only has a freer play in science than elsewhere, but can produce as great effects upon the world through science as through politics, philosophy, or literature, there will be more Darwins. Such men are interested primarily in truth as such, but they can hardly be quite uninterested in what will happen when they throw down their dragons teeth in the world.' 'The scientific worker of the future will more and more resemble the lonely figure of Daedalus as he becomes conscious of his ghastly mission, and proud of it.

"Black is his robe from top to toe
His flesh is white and warm below,
All through his silent veins flow free
Hunger and thirst and venery,
But in his eyes a still small flame

> Like the first cell from which he came
> Burns round and luminous, as he rides
> Singing my song of deicides."'

So much for Haldane's *Daedalus* of 1924. If Haldane had lived in 1994, one might ask, what would have been his predictions?

Surely he would not have been as optimistic as he had been in his youth. He would have seen the dangers of world overpopulation, the ever increasing pollution, the destruction of the natural environment, and whatever other ominous developments one can perceive in the world of today. On the other hand, he would have been pleased to see that biology had indeed become the queen of the sciences, as he had predicted. He would have been particularly gratified over the rise of molecular biology and its magnificent achievements. Although he was present at the beginnings of this development (he died in 1964), he was not alive to see the developments of genetic engineering and the link between molecular and evolutionary biology.

He would have probably done relatively little speculating with respect to physics and engineering. Indeed, he would have been amazed that man had been able to visit the moon and at the unbelievable development of computer technology and its applications. He would have realized the potential of genetic engineering, but would have seen its limits more clearly than some of the current enthusiasts. There are no methods in sight that would permit the engineering of highly complex genotypes.

Instead he would have thought a good deal about developments relating to human health. The success of the antibiotic revolution led to a shift of interest from infectious diseases to malignant and degenerative diseases. I think Haldane would have seen that the ultimate conquest even of those diseases would not lead to a lengthening of the human life span, but only that more people would approach the normal ultimate life span. Owing to his great knowledge of biochemistry, he might have been able to suggest why certain forms of cancer, that of the stomach, for instance, have decreased in frequency while others, such as cancer of the brain, have increased to such an extent.

In 1924 he had speculated about the popularity and benefits of stimulants, but by now he would have become deeply interested in nutrition. The finding that poorly nourished rats have a much greater life expectancy than well nourished ones would have induced him to suggest that experiments be done raising human infants on a reduced diet and following their health throughout their lives. Indeed, he would have become very much interested in the effect of different diets on human health and on susceptibility to various diseases and malignancies. He might have suggested placing the inmates of various penitentiaries on different diets (none of them, of course, deleterious) and studying the effects of these differences on their health. This would have included research on the effect of supplementary vitamins and minerals. Indeed, he might well have proclaimed that the study of human nutrition was one of the most neglected fields of preventive medicine.

Haldane was very much interested in education. In a splendid short essay (1949) he pointed out the difference between equality and genetic identity. As a good Darwinian he was fully aware of the genetic uniqueness of every human individual and, as a result, of the different capacities of different human beings. In order to give all of them equal opportunities, as demanded by the principle of equality, it is necessary to provide different individuals with different environments. Hence, contrary to the beliefs of some defenders of democracy, children with different capacities have to be placed in different schools. With eugenics having clearly been shown to be impossible at the present time, he would have stressed the need to do more for bringing out the most from every genotype, that is, a stress on the development of the best possible phenotype of every genotype. What particular changes in the environment both in respect to nutrition and other aspects of lifestyle as well as education would be necessary to achieve such optimal phenotypes would have to be one of the first priorities of future research.

I am quite sure he would have placed great emphasis on education for the development of improved value systems. At the time of our grandparents in the education of children there was a great deal of emphasis on the adoption of high morality. All the readings of the young child were stories with a moral. In the religious education morality played a great role, as indicated for instance by the term 'Protestant ethics'. All this has largely disappeared and the mass media make no effort whatsoever to contribute to the acquisition of high value systems by children and young people. I am sure Haldane would have singled this out for special attention. He might have gone so far as to demand that moral education was made a definite subject to be taught in schools. In the past much of this, of course, has been considered to be the duty of the parents. But since in certain segments of our society the normal family structure no longer exists, such parental education has become difficult in those segments. The high juvenile crime rate is a clear indication of a breakdown of moral education, and I am sure Haldane, who forever was a great believer in the improvement of human society, would have made proposals for the elimination of this weakness.

I have no idea what he would have proposed for the implementation of these ideas. Although he eventually left the communist party when he learned of the magnitude of Stalin's crimes, he probably continued to be a believer in the basic Marxist ideals. But of course, even those have become dubious after the complete breakdown of almost all Marxist regimes throughout the world. The great masses are simply incapable of exercising (without reward) as much altruism as the Marxist system demands. I have no idea how Haldane would have met this problem. That much is certain, that he would have thought about these problems deeply and with the great warmth of his heart.

REFERENCES

Adams, M.B. (1990). *The wellborn science: eugenics in Germany, France, Brazil and Russia*. Oxford University Press, New York.

Haldane, J.B.S. (1932). *The causes of evolution*. Longmans, Green, London.

Haldane, J.B.S. (1949). Human evolution: past and future. In *Genetics, paleontology and evolution* (ed. G.L. Jepsen, G.G. Simpson and E. Mayr), pp. 405–18. Princeton University Press.

Huxley, A. (1932). *Brave new world*. Penguin, New York.

Mayr, E. (1982). *The growth of biological thought*, p. 624. Harvard University Press, Cambridge, MA.

CHAPTER SEVEN

THE PARALLEL LIVES OF H.J. MULLER AND J.B.S. HALDANE—GENETICISTS, EUGENISTS, AND FUTURISTS

ELOF AXEL CARLSON

Like Castor and Pollux, Haldane and Muller resemble twins, more bound to each other by their similarities than set apart by their differences (Carlson 1982; Clark 1968; Dronamraju 1985). Their careers were roughly alike, both were primarily geneticists, often contributing to the same fields of human genetics and evolution; both were socialists (self-proclaimed Bolsheviks); both were eventually disillusioned especially by the rise of Lysenkoism and the state destruction of science as they knew it (although it took Haldane almost 15 years longer to lose his faith in the USSR); both were atheists and materialists in their world-view; both had a wide range of interests and talents outside their best known profession; both did not hesitate to speak out on social issues and the applications of science to society; both were intensely curious and speculated freely; both left their countries in disappointment; both taught or lived in several countries; both were recognized by their peers as geniuses.

They also differed. Haldane was raised in a patrician or aristocratic tradition. Muller was raised by his mother in genteel poverty through most of his childhood after the death of his father, and in his youth he experienced hunger, the lack of money to enjoy a full time undergraduate and graduate education, and an imposed frugality that left its mark long after he became relatively affluent as a successful scientist. Haldane was often rude, loud, and abrasive in his relations with others. Muller was socially restrained and shy; he allowed his discontents to fester inwardly into sullen, life-long dislikes of those he believed had let him down. Haldane enjoyed poetry and the classics and had sympathy for the ultimate strivings (but not the findings) of religion. Muller was uncomfortable with the arts and humanities, his taste in literature limited to science-fiction, and he had little respect for religion except in the guise of Humanism. Haldane felt secure in his career, his opinions, and his personality, and he enjoyed his eccentricities. Muller was insecure, competitive, irritable, riddled with self-doubts, and on at least one occasion suicidal. Haldane was a gifted mathematician; Muller was a gifted experimentalist. Haldane enjoyed popularizing science to the general public through the popular press; Muller preferred to write for a broader audience of scientists and rarely wrote for the general public.

Haldane gives us a view of 'science and the future' the subtitle of his *Daedalus*. His audience is a college club, the Heretics, at Cambridge and he presents his unleashed imagination in 1923. It is an audience that was familiar with Jules Verne and H.G. Wells and thus science fiction or the ways science can transform society was not a shock. But Haldane did shock his audience and the readers of his slim little volume when it was published the following year (Haldane 1924). His future vision shocks because Haldane raises for the first time something that surpasses the much tamer ambitions of eugenists in England and America of that era. He applies experimental embryology and advances in reproductive technology in animals to a not too distant human future of some 150 years. He offers a control over reproduction that far surpasses birth control. It permits the separation of the act of sexual intercourse from the process of reproduction. He invites his audience to consider 'ectogenesis', the gestation of an embryo from fertilization to birth outside the uterus and the anticipation of what would be called 'test tube babies'. Yet Haldane claims his anticipations of science applied to humanity are 'modest, conservative, and unimaginative' (p. 26). Some are. He sees a future where food can be synthesized and agriculture as we know it dispensed in favour of sugars and starches from cellulose; there would also be amino acid soups from a chemical manipulation of coal and atmospheric nitrogen. He dismisses eugenics not for its potential to change humanity but for its present dull and limited (if not erroneous) application of genetic principles. He identifies 'the eugenic official' as a composite of 'the policeman, the priest, and the procurer. . .' with one's eugenic partner chosen by 'something of the nature of a glorified medical board' (p. 35). He prefers the ectogenic propagation of large numbers of children utilizing the desirable genetic traits of established gametic stocks. His eugenic values are largely unexplored. He reveals (p. 42) his own implied social range of hereditary traits in telling his audience that such a planned breeding *in vitro* would provide a world with benefits 'from the increased output of first class music to the decreased convictions from theft. . . .' His focus on human reproduction is remarkably brief, tucked in the last few pages of an undergraduate's report that he imagines will be written midway in the twenty-first century (pp. 39–43). Most of that student's future essay is devoted to the consequences of turning the oceans purple with a genetically engineered bacterium that shifts the food chain (and our diet) from land to sea. He does not hesitate to invoke, using only a few selected gamete donors for the ectogenic propagation of most of humanity, what had to strike his audience and readers (such as Aldous Huxley) as a radically different, if not chilling, world.

Muller had a more intense commitment to eugenics than Haldane. He advocated eugenic views from his undergraduate days at Columbia College in 1910 in an address to a debating club, the Peithologian Society. He extended his ideas while at the University of Texas in 1925 and completed his popularization, *Out of the night*, while in the Soviet Union as a Guest Investigator carrying out an active programme of basic research on radiation genetics and gene structure (Muller 1936). Muller claims in his preface that the Peithologian address and an expansion of that talk in 1925 provided the bulk of his book. His 1931 and 1935 revisions are largely expansions of his socialist ideals and updates on eugenic technology. Muller's eugenic views are primarily Galtonian. He favoured positive eugenics, the breeding of the best traits to benefit the next and future generations. He took dim interest in the American eugenics movement led by C.B. Davenport, H.H. Laughlin, and associates of the Eugenics Record Office at Cold Spring Harbor, NY, especially as that movement targeted social classes, ethnic groups, and races for genetic inferiority and backed compulsory sterilization laws and restrictive immigration policies as its means of applying eugenics to society (Kevles 1985; Reilly 1991).

In contrast to Haldane's brief mention of genetics, Muller spends the greater part of his book (more modestly subtitled 'a biologist's view of the future') on the centrality of the gene in evolution, its determination of our material biology, and the way in which inherent differences in human abilities and behaviour can be teased apart from the complex overlay of environmental influences, cultural traditions, training, and education that shape cultures and individuals in such different ways (Muller 1929). Where Haldane uses the capsule history of his imaginary student's report to create surprise and to disturb the complacency of his audience, Muller builds a non-fiction utopia in which science gradually changes the world for the good of humanity. We are introduced to feminism, absent in Haldane's account of the future: 'Only by lightening the physiological, the psychological, the economic, and the social burdens on the mother now caused by child-bearing and child-rearing can we attain to a state in which eugenics is feasible' (p. 129). Muller stresses the importance of the woman's contribution to heredity and why women must be given opportunities, like men, for professions and creative work to reveal their genetic talents. Haldane's *Daedalus* is characteristically masculine, reflecting that generation's men's club mentality in educated British society. Muller's commitment to feminism stems from his first wife, Jessie Jacobs Muller, who was a Ph.D. mathematician who was fired from the University of Texas faculty on the birth of their son, David. Her plight left a lasting impression on Muller of the irrational prejudice against women who sought both a family and a career.

Muller also steers away from the formidable problems of ectogenesis and favours the use of cultured sperm (preferably kept 25 years until after the death of the donor to permit a more objective evaluation of that person's worth, an idea that he retrieves some 20 years later when frozen sperm are introduced into livestock breeding). He foresees some 250 000 inseminations from each voluntarily chosen male donor and the process repeated with each generation's best endowed individuals in intelligence and personality. Muller insists that selection for high intelligence is insufficient and the great advances of civilization depend on the cooperative, caring, sharing, and positive personalities that make social equality and justice possible. He recognizes that the sex preferences of his day are biased towards males but does not object even to the physical separation of male and female determining sperm. The initial abuse in favour of male children by ordinary men and women electing this option would eventually balance out. Those men attracted to the values of eugenics would be willing to choose a sperm donor of eugenic worth if their second child (he calls them 'children of choice') were a daughter. This would reflect the same bias as infertile couples who prefer females for adoption. Muller also notes that sexual selection could be used to prevent X-linked disorders in a relatively few generations.

In Muller's eugenic world, each generation would be benefited by more than a hundred-fold increase in eminent men and women. Repetition of the process over some 200 years would lead to a world of talented and gregarious people whose accomplishments would far exceed what we could imagine from the slow pace of civilization over the past 7000 years. If Haldane's view of the future is that of science overriding any reflective social thought on the just or desired society, Muller's future is one in which the control of human evolution is taken away from nature and placed in our own hands. He develops almost a religious fervour in describing that future world: 'In time to come, the best thought of the race will necessarily be focussed on the problems of evolution—not of the evolution gone by, but of the evolution still to come—and on the working out of genetic methods, eugenic ideals, yes, on the invention of new characteristics, organs, and biological systems that will work out to further the interests, the happiness, the glory of the god-like beings whose meager foreshadowing we present ailing creatures are' (p. 156).

Muller has often been taken to task by critics of eugenics for his 1936 choices of the calibre of men whose sperm he would like to see propagated on a large scale (Haldane was wise not to choose among past or present figures). Muller suggests '. . . it would be possible for the majority of the population to become of the innate quality of such men as Lenin, Newton, Leonardo, Pasteur, Beethoven, Omar Khayyam, Pushkin, Sun Yat Sen, Marx (I purposely mention men of different fields and races), or even to possess their varied faculties combined' (p. 141). In addition to the Asian and Middle East heritage, Muller used Pushkin who was believed to have an African ancestor. Muller's choices of talent are science (3), art (1), music (1), literature (2), and political leadership (3). Usually Muller is quoted for this pantheon of potential sperm donors without the parenthetical statement or the combinatorial bonus of pooled talents. His choice of Lenin and Marx (especially during the Cold War) was raised as an example of Muller's bad judgment. Muller tried to defend himself by talking about the qualities of these men—for Lenin and Marx it would be their creativity, their leadership, and their ideals of a society that abolished discrimination based on class. Muller never acknowledged that he, like most of his contemporaries, could be influenced by the popularity, the ideology, and the cultural beliefs associated with the worth of political figures. Even his 25 year waiting period would not have rescued him from the error of using

Lenin. Lenin is a dubious choice (using Muller's own criteria) because he may have been insensitive to the lives of his opponents and allowed (or ordered) their deaths to protect his own power as head of state. Muller also did not recognize that whatever talent for leadership and caring about society may reside in the genes, it may manifest itself in patriotism or ideological loyalty that could be evil in its consequences to others.

If Muller's genetic assumptions about the heritability of intelligence, creativity, artistic talent, leadership, gregariousness, and caring are valid, his choices are appropriate except for the gregariousness and caring that would be hard to assign to his political trio or to any of his scientific or cultural heroes. Muller did argue that it is the potential for, not the repetition of, a specific eminence that he sought to propagate. The nurturing values Muller admired are often manifest in religious innovators. We think of Jesus, Buddha, St Francis, or in modern times Albert Schweitzer and Prince Kropotkin, as people who preached a message of love, cooperation, and caring. Muller shied away from these exemplars because they were usually committed to a religious life. Kropotkin was from the overthrown Russian aristocracy and royalty and this may have inhibited Muller from using him.

Much more serious for both Haldane's and Muller's eugenic world of superior people is the absence of compelling evidence for the heritability of the qualities that they advocate. We do not know whether the association of bright children with bright parents is primarily genetic or primarily owing to a family's social influence. We have no evidence that meanness or kindness is primarily genetic. In the absence of such evidence any positive eugenic scheme is premature and may not be necessary. If social influence so profoundly creates differences in values, speech, and outlook among different cultures, classes, professions, religions, and political outlooks, to what degree would genetic components for behaviour override these? How would one select for sperm donation between a suitably behaved person trained by his culture and a genetically endowed male whose personality showed the same behaviour?

As the passage of time permits a more detached view of Muller and Haldane, we can assess how their personalities were shaped by their times. Both men shared a view of progress that was a heritage of the Enlightenment. From Condorcet came a belief in the unlimited potential of science to understand all aspects of the universe and of reason to solve all human problems and lead humanity out of its bondage to kings and tyrannies. From Goethe we obtained the Faustian personality who serves as a model for the genius who never stops learning and who finds redemption for his anti-establishment acts (like a Prometheus or Daedalus in antiquity) by applying science to human benefit. It is the Faust of book II who drains the swamps, develops harbours, and reclaims the once pestilential land for agriculture, thereby improving the lives of hundreds of thousands of suffering humans. This combination of the hero and the notion of progress is a hidden assumption in the eugenic worlds of Haldane and Muller. The individual who changes society, who takes charge and uses his talents and creativity to bestow benefits and culture to humanity is the model used for these two eugenic systems. It was also the model used in a perverted way to justify the robber barons, like Andrew Carnegie, whose patron saint was Herbert Spencer, whose philosophical underpinning was social Darwinism, and whose future outlook was 'the Gospel of Wealth', with its dispensation of immense profits for the establishment of colleges, libraries, and learned institutions to better humanity. It is also the model, in an even more perverted way, of a Hitler or a Mussolini, men of supreme confidence in the good they were doing for society even if it meant purging a sizeable proportion of innocent persons who stood in their way.

Haldane is more philosophical than Muller in revealing his scientific values. Muller sticks closely to a socialist ideal of a humanity free of racial, sexual, or class prejudice. Haldane sees scientists and scholars as heretics and his choice of Daedalus as the patron saint of scientists is deliberate: 'These are the wreckers of outworn empires and civilizations, doubters, disintegrators, deicides.' Among his 'wreckers' he includes Voltaire, Bentham, Thales, Marx, and Darwin (p. 45). While he does not offer his heroes as exemplars to be used as sperm donors, it is clear that the same qualities admired by Haldane in his heroes were also admired by Muller.

Haldane paid passing acknowledgement to eugenics but did not go to any lengths to promote it in his later years. Muller made a personal crusade out of eugenics and only fear for his career (in the 1920s), the mockery of eugenics by Nazism, and the shock of genocide revealed by the Holocaust made Muller push his eugenic agenda to a back burner. He revived it in the mid-1950s and hoped to breathe a rekindled interest from the embers of a dying if not dead eugenics. His stress on voluntarism in 'germinal choice' found no takers among those he hoped to reach. His last hope, a sperm bank for those with the qualities of intelligence or caring personalities, he abandoned when its financial backers revealed values too close to the American eugenics movement for him to accept. The Foundation for Germinal Choice went to the grave with Muller but a sperm bank for geniuses (essentially defined by high SAT or GRE scores because of the paucity of donors among Nobelists and other public figures) is still active in California providing modest amounts of children who may turn out to be the next generation of 'Terman children' (Terman 1925). Lewis Terman studied the lives and careers of high IQ children for nearly two generations. Unlike eminent individuals about whom biographies are written, the high IQ child or the high GRE or SAT score student is not usually the person who finds distinction as a public figure. Those geniuses whose works we admire are more often identified through their creativity, their energy, their ability to see things to completion, and their intense commitment to an idea or a career. Their personalities may be more tortured and their family life more stressed than those of the smooth-sailing 'Terman kids' (Goertzel 1962). None of these aspects can be measured well, if at all, by tests of academic competence.

There will be a time to take stock of the genetic differences among talented individuals with diverse personalities in the none too distant future. The advances in neurobiology over the past 20 years, the use of the human genome project early in the twenty-first century, and the likelihood that functions such as memory, problem solving, and relational thinking will be resolved to a neuronal level could provide a more substantial basis for a genetics of human behavioural traits (Crick 1994). If there is a significant genetic component to these traits, many opportunities will become available for sperm and egg donors to provide gametes for infertile couples (as much as 10 per cent of humanity) or persons who might be persuaded by Muller's eugenic values to have at least one of their children conceived as a eugenic choice. It is quite possible that the search for a genetic basis for talents of social worth may be an illusion. The neurobiology of the future may reveal a far greater plasticity of personality and potential for achievement (i.e. the capacity to learn) for most of humanity than we presently admit and the focus of such a future society will not be on positive eugenics but on environmental programmes that bring forth the talents latent in most of our children.

Both Haldane and Muller rejected negative eugenics. Their distaste was based on compelling grounds. The programmes then popular were racist, sexist, and class based. The instruments to measure social failure were slums, prisons, asylums, and the homeless in the streets. Those advocating a negative eugenics programme offered containment by restrictive immigration, isolation through marriage laws and pariah status, and surgical sterilization. All of these approaches were coercive, grounded in spurious assumptions about the reasons for social failure, and genetically ineffective except over many centuries if the assumed recessive genes for these anti-social traits did exist.

At that time little was known about genetic disorders because almost all of those children born with such defects died shortly after birth or were lumped together in generic categories such as multiple malformations or failure to thrive. Today there is an immense literature on some 4000 monogenic traits in humans and several hundred can be diagnosed *in utero* by amniocentesis, ultrasound, and other techniques. Except for the dominant traits and the X-linked traits, amniocentesis with elective abortion does not diminish the incidence of the genes involved because most people having the conditions for which elective abortion is chosen are so sick or compromised in function that assortative mating would prevent their having children if they reached reproductive maturity. Assortative mating is the outcome of the conscious or unconscious reasons we use in selecting our mates. College graduates rarely marry the mentally retarded; most individuals with incapacitating, chronic, or fatal illnesses do not find partners for marriage; most marriages involve the same race, religion, and social class upbringing. Many of the traits involved are not genetic, but many others do have strong genetic components.

As Haldane (and independently Fisher) first noted, the mutant gene is overwhelmingly likely to be found among normal-looking carriers in the population rather than among those infants expressing the recessive trait. As an example, if a recessive genetic disorder, such as albinism, is born once among 10 000 children, the frequency of carriers, like the parents, is about 2 per cent of the population. The attempt to reduce substantially this gene frequency in the population by sterilization of albinos would take millennia.

Negative eugenics would only work effectively if carriers as well as those receiving a double dose (homozygosity) of the recessive mutant gene were aborted or prevented from coming into being as zygotes. It would also work if one of the two carriers at risk (preferably the male) chose artificial insemination from a donor who was screened and shown not to carry the recessive gene. Muller and Haldane both recognized that new mutations and pre-existing mutations in the population are not as readily selected out under the conditions of modern culture with its public health programmes, greater nutrition, and social services. How much of an increase in the amount of heterozygous mutant genes the population can carry without expressing a variety of illnesses and other burdens is not known.

Critics of eugenics have argued against it mostly on political grounds (it leads to Holocausts), ethical grounds (it is artificial and diminishes our sense of humanity), religious grounds (it is playing God), and philosophical grounds (it reduces variety or some hoped for benefit of being heterozygous for any given gene). They do not address the issue of where new mutations will end up, the usually deleterious nature of gene mutations, the misery assigned to individuals fated to receiving genes leading to medical syndromes, the psychological burdens on most parents who raise such children, and the financial costs borne by families or society in paying for health needs, educational needs, and social needs of such children. Nor do many of these critics consider it appropriate for parents to choose the quality of life of the children they wish to bring into the world.

Eugenics today is largely operating through unintended changes in society. Assortative mating keeps those with serious genetic difficulties from finding a marital partner. Co-educational rather than single sex colleges bring together bright relatively healthy students who marry each other. Prenatal diagnosis, genetic screening, artificial insemination, and *in vitro* fertilization do provide some negative eugenic outcomes and have a potential to do so on a larger scale as the analysis of genes in the human genome project reaches a peak in the early twenty-first century. These expectations are very different from those of only a few generations ago. One could argue that at one time almost all marriages were arranged by parents; at one time almost all conceptions led to birth or spontaneous abortion; at one time almost all acts of sexual intercourse were unprotected for their reproductive outcomes. All three of these have largely been replaced by a greater respect for individual values, changes in the laws and practices of society, and the technology to carry out human needs.

So far the negative eugenic consequences of life in our times have been fortuitous and not intentional. Most physicians, genetic counsellors, and clients seeking screening or prenatal diagnosis are not concerned with human evolution, the idea of a gene pool, or the improvement of humanity. They seek help to prevent the birth of a child who will experience pain or diminished participation in life; they also do it for themselves as parents who want a 'healthy' child. Their values are pragmatic, not idealistic: 'Why should I bring a sick child into the world when I can use medical technology to bring a healthy child to raise?' We are ambivalent as a society in our responses to children born with birth defects and other genetic conditions that depart from societal norms (Carlson 1992).

Both Muller and Haldane knew that in the absence of any eugenic programme and assuming the validity of their assumption that civilization prevents effective natural selection, the load of mutations in humanity would gradually rise. No serious crisis would result for a millennium or more of such neglect. A millennium is a long time and no one can predict now the values of that future generation, the new technologies for preventing zygotic unions of genes that would impair a child's health, or the insights into our nervous system, behavioural origins, or developmental processes. Neither the implementation of a positive or negative eugenic programme on a modest scale nor the absence of such programmes in the immediate future is likely to have much impact on the world's gene pool over the next thousand years. Nor need we fear some universal compulsory sterilization programme, universal ectogenesis, nor a universal forced mating system such as we encounter in science fiction dystopias. We must remind ourselves that 'Thousand Year Reichs' last only 13 years!

REFERENCES

Carlson, E. (1982). *Genes, radiation, and society: the life and work of H.J. Muller*. Cornell University Press, Ithaca, NY.

Carlson, E. (1992). 'Human imperfection: unresolved responses'. *Quarterly Review of Biology*, **67**, 337–41.

Clark, R. (1968). *JBS: the life and work of JBS Haldane*. Hodder and Stoughton, London.

Crick, F. (1994). *The astonishing hypothesis: the scientific search for the soul*. Scribner's, New York.

Dronamraju, K. (1985). *The life and work of JBS Haldane with special reference to India*. Aberdeen University Press.

Goertzel, V. and M. (1962). *Cradles of eminence*. Little Brown, Boston, MA.

Haldane, J.B.S. (1924). *Daedalus: or science and the future*. Dutton, New York.

Kevles, D. (1985). *In the name of eugenics: genetics and the uses of human heredity*. Alfred Knopf, New York.

Muller, H.J. (1929). *The gene as the basis of life*, Proceedings of the International Congress of Plant Sciences, 1926, Vol. 1, pp. 897–921. Ithaca, New York.

Muller, H.J. (1936). *Out of the night: a biologist's view of the future*. Vanguard Press, New York.

Reilly, P. (1991). *The surgical solution: a history of involuntary sterilization in the United States*. Johns Hopkins University Press, Baltimore, MD.

Terman, L. (1925). *Genetic studies of genius*. Stanford University Press.

CHAPTER EIGHT

Daedalus, Haldane, and Medical Science

D.J. Weatherall

At first sight it is not clear why J.B.S. Haldane chose Daedalus, the legendary Athenian craftsman, as the hero of a paper on the future of science, read to the Heretics at Cambridge in 1923. For, as he points out, apart from Daedalus's ability to construct statues which could move themselves, and the invention of glue, we are told very little about his achievements.

But the story of Daedalus, if fragmentary, is certainly not uneventful. Like many teachers, it appears that he was happy for his students to do well, but not too well. Because of the fear that his nephew and pupil, Talus, who is said to have invented the saw and potters wheel, might outshine him, he threw him down from the Acropolis. For this ungenerous act he was condemned by the Areopagus and fled to Crete, where he built the Labyrinth for King Minos. It was here that the Minotaur was housed, and fed on a diet of Athenian youths and maidens. To prevent him leaving Crete, Minos confined Daedalus to the maze, together with his son Icarus. Clearly it was time to move on again, and so he constructed a pair of wings from wax and feathers for himself and his son, and flew away. But Icarus, because he ventured so near to the sun that the wax melted, fell into the sea and was drowned. Daedalus, who was not involved in the first recorded air disaster, escaped to Sicily, where Minos, pursuing him, also seems to have met with a violent death. What eventually happened to Daedalus is not recorded.

It appears that it was the very fact that Daedalus sank into obscurity that made him so attractive to Haldane; 'the most monstrous and unnatural action in all human legend remained unpunished in this world, or the next.' He was, suggests Haldane, the first to demonstrate that a scientist is not concerned with the Gods. 'But if he escaped the vengeance of the Gods he has been exposed to the universal and age-long reprobation of a humanity to whom biological inventions are abhorrent. . . .'

It was by this rather circuitous route that Haldane introduced his Cambridge audience to the notion that the really surprising scientific developments of the future would be in the fields of biology and medicine rather than chemistry or physics. By inviting them to listen to some extracts from an essay on the influence of biology on history during the twentieth century, written by 'a rather stupid undergraduate member of this university for his supervisor during his first term 150 years hence', he painted a disturbing picture of ectogenesis, eugenics, the control of the less agreeable aspects of ageing, and at least by inference, genetic manipulation. Both in the introduction to *Daedalus*, and in the run-up to the section on the biology of the future, it is clear that Haldane expected his audience to find much of what he had to say unpalatable, or worse.

To what extent are Haldane's predictions for the future of human biology in general, and medical practice in particular, likely to turn out to be correct? How far have we already moved towards his world, and what might happen in the 79 years that remain before the year that Haldane's imaginary student presented his essay? Haldane's writings on this subject were not, of course, confined to *Daedalus*, and in later years he was to develop many other highly original ideas in the fields of medicine and biology which, in the event, have had a much more profound influence than *Daedalus*. But *Daedalus* does have some interesting and important messages for today, particularly with respect to its ethical implications for a society which is having serious doubts about the direction in which the biological sciences are moving.

HALDANE'S BRAVE NEW WORLD

It is likely that audiences in the early part of this century would have found Haldane's predictions about ectogenesis the most disturbing part of his essay. They were adapted with horrifying effect by Aldous Huxley and George Orwell, and are still to be found in current works of science fiction. Haldane predicted that the first ectogenetic child would be produced by 1951. This achievement would follow the recovery of a fresh ovary from a woman who was the victim of a more conventional flying accident than Icarus. The idea would be to maintain the ovary alive in culture medium for several years, during which time eggs could be obtained and fertilized successfully. Within a few years it would be possible to take an ovary from a woman and maintain it in tissue culture for as long as 20 years, producing a fresh ovum each month, of which 90 per cent would be fertilized. The embryos would then be grown in incubators for nine months. By 1968 the French would be producing 60 000 children annually by this method and, although there would be some opposition, notably from Rome, by the middle of the twenty-first century less than 30 per cent of children would be born of women. (We are left to guess why the French were chosen for this particular success story.)

Although Haldane admitted that this method of human reproduction might have some deleterious effects on family life, he believed that it would become generally accepted that the beneficial effects of selection would more than counterbalance these minor social inconveniences. The small proportion of men and women who were selected as ancestors for future generations would be so undoubtedly 'superior to the average' that the advance in each generation in any single respect, from an increased output of first class music to decreased convictions for theft, would be quite startling. Had it not been for ectogenesis, he suggests, civilization would have collapsed because of the greater fertility of the less desirable members of the population in almost every country.

But this would only be the beginning. By 1990 biological manipulation would have become even more radical. Reproduction would be separated completely from sexual love, and mankind would be 'free' in a new sense of the word. In essence, it would be possible, by selective breeding, to change character as quickly as institutions. The election placards of 300 years hence will read 'Vote for Smith and more musicians', or 'Vote for McPherson and a prehensile tail for your great-grandchildren'. Haldane reminds us that it has been possible to alter animal species to a remarkable extent and hence it seems only a question of time before it will be feasible to apply the same principles to Man. We will be able to control our passions by more attractive methods than fasting and flagellation, and to stimulate our imagination by reagents with less after-effects than alcohol; in short, we shall learn to deal with our perverted instincts by physiology rather than prison. Running through all these predictions there is the belief that, for all the problems that advances in biology may bring with them, the overall eugenic effects of our increasing sophistication in biological manipulation will be of enormous benefit.

After these extraordinary predictions *Daedalus* rather peters out with respect to the future of medicine, ethics, and religion. Disease will only be conquered if we turn from the laboratory sciences of Pasteur and Koch to whole-body physiology. As illness is controlled, death will become a more straightforward physiological affair, like sleep. By artificially preventing the menopause women will grow older 'as gracefully as men' and concerns about an after-life will decline; Haldane's rather strange views about women, which recur several times in *Daedalus*, were—and to some extent still are—not unusual for products of the monastic isolation of the British public school system. More surprisingly, spiritualism might just achieve scientific verification and hence become a major competitor with the great religions. But we must learn not to take traditional morals too seriously. Because even the least dogmatic of religions tends to associate itself with some kind of unalterable moral tradition there can be no truce between science and religion. Even the more adaptable ones, such as Hinduism and Christianity, will not admit that their mythology and morals are provisional, the only sort of religion that would satisfy the scientific mind and which, Haldane suggests, could hardly be called religions at all. In short, the scientist of the future will come more and more to resemble 'the lonely figure of Daedalus as he becomes conscious of his ghastly mission, and proud of it'.

HOW FAR HAVE WE MOVED TOWARDS HALDANE'S WORLD?

With close on 80 years to go it is interesting to reflect on how far the biological sciences have advanced towards realizing Haldane's predictions. The production of 'test-tube babies', or genuine ectogenesis as defined by Haldane, is a little further advanced than it was when Haldane wrote his essay. For despite the remarkable advances in the management of extremely small premature infants, it has not been possible to develop animals or humans from fertile eggs outside the body of their mothers. On the other hand, *in vitro* fertilization, with the return of the fertilized eggs to the mother, is now routine clinical practice and is of great value for the management of some forms of infertility. Once this technology was established it was soon realised that fertilized eggs need not be returned to the biological mother and that it would be feasible for other women to act as surrogates. Furthermore, by appropriate hormone treatment it has been possible for women in their late 50s, or even older, to bear children. And the advent of *in vitro* fertilization allows women to carry children of different racial groups to themselves; several black women have, for various social reasons, opted to bear white children after *in vitro* fertilization of ova obtained from other women. Donation of ova has, like donation of sperm, become commonplace. But because this has led to a shortage of ova, techniques are being developed to obtain them from aborted fetuses; offspring produced by *in vitro* fertilization of eggs from this source would be in the remarkable position of being born of a union in which one parent had never existed beyond the fetal stage. And in the near future it is likely that ovaries, like other organs, will be removed from cadavers, in this case as a source of eggs for *in vitro* fertilization.

Although there are hints of the possibility of genetic manipulation in *Daedalus*, even Haldane might have been surprised at the extraordinary developments which have followed the advent of molecular biology over the last 40 years. The ability to isolate and transfer single genes has opened up a remarkable new world, in particular the development of transgenic animals, that is animals which express 'foreign' genes that have been injected into fertilized ova. A wide variety of human genes has been expressed in mice and larger animals and stably transmitted through many generations. Although transgenic experiments of this type have not been carried out in human beings the technology is sufficiently advanced that it may be possible. The products of human genes, expressed in the milk of transgenic animals, are already being developed as therapeutic agents. And the first faltering steps at somatic gene therapy, that is the transfer of genes directly into cells other than germ cells, have been taken. It now seems certain that several single gene disorders will be corrected in this way, and long before AD 2075.

Although Haldane undoubtedly anticipated the advent of diagnostic genetics in his later writings, he did not touch on this subject in *Daedalus*. It is now possible to identify the carrier states for many serious single gene disorders and to diagnose affected fetuses as early as nine weeks of gestation, or even in ova after *in vitro* fertilization. Termination of pregnancies for severe congenital malformation or inherited disease is commonplace.

It seems certain, therefore, that well before Haldane's deadline, screening for many common single gene disorders will be available and families will have the choice of whether or not to bring a child with a serious disease of this kind into the world. Somatic gene therapy, either by replacement of a defective gene or by correcting it by site-directed recombination, nature's way of swapping parts of genes, will be available for some of these diseases, and society may or may not have accepted the availability of germline gene therapy, that is the correction of monogenic diseases by injection of appropriate genes into fertilized eggs. Using this approach, of course, the 'foreign' genes would be expressed in the offspring of the recipient and hence we would be altering the genetic make-up of future generations. It also is likely that the genes involved in making us more or less susceptible to some common acquired disorders such as heart disease, stroke, diabetes, the major psychoses, Alzheimer's disease, and others, will have been identified. While medical advances may make it possible to use this information to prevent or treat these diseases, we may go through a long period when we may have this information and do not know what to do with it. There is no doubt that many human genes will be used commercially to underpin an expanding biotechnology industry, for the production of a wide range of proteins and other therapeutic agents. And transgenic animals carrying human genes will be used increasingly for the same purpose.

Haldane's thoughts on ectogenesis may well have led to the notion, so beloved by later novelists and science fiction writers, of cloning human beings, that is producing lines of genetically identical people. Although genetically identical adult vertebrates, both normal and fertile, were first produced in 1961 by transplanting nuclei from an embryo into a number of enucleated recipient eggs of *Xenopus*, despite rumours to the contrary it is highly unlikely that an experiment of this type has been carried out in human beings. On the other hand, our ability to isolate, characterize, and manipulate human genes has, in the field of human genetics, given us the potential to fulfil some of Haldane's ideas about altering our genetic constitution. In principle, if not in detail, his forecast that, by the early part of the next millennium, we would have gained a significant control over our futures has been borne out. And, in passing, so has an even more remarkable prediction that Haldane made at about the same time.

One of the major goals on which so many hopes for developments in human biology and medicine over the next half century are pinned is called the Human Genome Project, that is constructing a complete linkage map and determining the nucleotide sequence of the DNA of all our chromosomes. The first stage in this endeavour, the construction of a genetic linkage map, is now well advanced. The advent of DNA technology led to the discovery of highly polymorphic regions of human and animal DNA which result in a considerable amount of genetic variability between individuals. Such mini- or microsatallite DNA has given us linkage markers which have proved invaluable for isolating the genes that are involved in important monogenic disorders and for their prenatal detection. Using this new kind of linkage analysis it has already been possible to obtain useful linkage maps of all the human and mouse chromosomes.

What is often forgotten by a generation which sometimes seems to believe that biology began with the discovery of DNA is that Haldane predicted the medical application of genetic linkage in an essay entitled *The future of biology*, first published in 1927, only three years after *Daedalus*. After pointing out that we know very little about human heredity, Haldane reminds us that several inherited characteristics have been found, such as the two which determine whether it is safe to transfuse blood, which are either present or absent, and admit no doubt. Furthermore, they are inherited in a very simple manner, and divide mankind into distinct classes. He goes on, 'now, if we had about 50 such characters, instead of two, we could use them, by a method worked out on flies by Morgan of New York and his associates, as landmarks for the study of such characters as musical ability, obesity, and bad temper. When a baby arrived we should have a physical examination and a blood analysis done on him, and say something like this: "he has got iso-agglutinin B and tyrosinase inhibitor J from his father, so it's 20/1 that he will get the main gene that determined his father's mathematical powers; but he's got Q4 from his mother, to judge from the bit of hair you gave me (it wasn't really enough), so it looks as though her father's inability to keep away from alcohol would crop up in him again; you must look out for that."'

This is a remarkable prediction which, I suspect, surpasses any that appear in *Daedalus*. For although some of Haldane's examples, bad temper and alcoholism for example, are unlikely to be due to the effects of a single gene, the principle which he was establishing, that is the use of genetic linkage to predict whether a baby might carry a particular genetic trait, has, of course, become one of the standard methods for prenatal diagnosis of genetic disease and, as we have seen, for obtaining a linkage map of the human genome.

Overall then, the young Haldane was not too far from the mark. We have made extraordinary advances in our ability to control reproduction and to avoid and even make a start at treating some of our more serious genetic diseases. Modern medicine has, to some extent, followed Haldane's advice and many of its major triumphs are the direct result of learning more about whole-body physiology. Modern high technology practice encompasses ingenious surgical advances and improvements in anaesthetic practice, organ transplantation, highly sophisticated methods for maintaining patients with multi-organ failure, the development of a vast and varied pharmacopoeia with the ability to alter almost any physiological function, and the rest. We have learnt how to eradicate or control at least some infectious diseases, and although we have made less progress towards the prevention or cure of the major problem diseases which took their place, there are few for which something cannot be done to alleviate suffering or prolong life. So where will the next 70 years take us?

HUMAN GENETICS AND MEDICAL PRACTICE OVER THE NEXT HALF CENTURY

Given these remarkable advances in biology and medicine, as we move to the close of the twentieth century we should be in a better position than Haldane to try to anticipate how medical practice will look by the middle of the twenty-first century. Assuming that society does not decide to put a major halt on biotechnology, and in its present mood of concern over the way the biological sciences are moving this cannot be ruled out, it seems likely that the medical scene will be quite different in 50 years time. But in looking to the future we need to be cautious; some of the claims of today's molecular biologists for the advances which may follow from the Human Genome Project make the writings of the youthful Haldane look very tame.

Many diseases that are entirely genetic in origin, that is single gene disorders and major chromosomal abnormalities, will be identifiable early in fetal life, and parents will have the choice of deciding whether to allow affected children to be born or to terminate the pregnancy. Although prenatal detection of disease may still be carried out early in pregnancy I suspect that many parents who are at risk for having children with this type of disease will opt for *in vitro* fertilization, detection of which fertilized eggs carry the defective gene, and replacement of those which do not. This will avoid the necessity of losing pregnancy after pregnancy, while waiting for one in which the fetus is not affected. Routine screening of pregnant women for monogenic disease or major chromosomal abnormalities of the fetus may, for reasons of cost, still be restricted to particularly common diseases. Thus babies with genetic diseases will continue to be born. But in 50 years time it is likely that some of them will be amenable to correction by gene transfer therapy, hopefully by some form of correction technology based on artificial recombination or the use of minichromosomes, rather than by the current 'hit and miss' methods of transferring genes attached to retroviruses. Thus although single gene disorders will still be with us, the scope for their avoidance and treatment will be greatly increased.

Despite Haldane's predictions, infectious disease will still be with us, and in 50 years time we will have witnessed several epidemics due to the emergence of 'new' infectious agents, like the epidemic of AIDS which is decimating societies in the developing world today. The control of infectious disease will remain a battle between the genetic cunning of germs and the ingenuity of the pharmaceutical industry. But because of the evolutionary efficiency of micro-organisms, new and virulent ones will always be turning up; they preceeded us by millions of years of adaptation and it is even possible that they may eventually see us off.

Some of the diseases that have replaced infection as the major killers in western societies, and which are starting to pose a serious problem in the developing countries, now that social conditions are improving and they are adopting the bad habits of the west, may still prove to be particularly intractable. They include heart disease, cancer, diabetes, the major psychoses, rheumatism, autoimmune disorders, and a variety of other chronic conditions which, though they can be controlled to varying degrees, are not easily preventable and cannot be cured. Remarkable advances in epidemiology have told us that many of them are likely to be due to the effects of environment and lifestyle. There is increasing evidence, however, that our genetic make-up renders us more or less susceptible to these environmental insults and that many of these diseases are also part of the complex processes of ageing. In short, the picture which is emerging is that many of our current ills reflect the effects of a completely new environment acting on a genome which has, over many generations, been adapted to completely different geographical and physical conditions and lifestyles.

This concept of the genesis of our current killers is based in part on twin studies which suggest that susceptibility to these diseases has a genetic component which varies in importance between them. For example, type 2 (insulin resistant) diabetes shows a concordance of nearly 100 per cent, whereas heart disease and hypertension, with rates of 20 to 30 per cent, have a much lower genetic component. Using genetic linkage analysis, and helped by information which is coming from the Human Genome Project, there is a major effort to discover the important genes for these polygenic disorders. Already major players in the generation of type 1 (insulin responsive) diabetes and other autoimmune diseases, type 2 diabetes, heart disease, Alzheimer's disease, osteoporosis, and premenopausal breast cancer have been found. It is likely that, by studying the action of these genes, and how it varies between susceptible and non-susceptible individuals, it will be possible to start to understand the pathogenesis of some of our most intractable killers and hence how better to prevent and treat them. And by the same token it may become feasible to identify subsets of our populations on which to concentrate our efforts at preventative medicine.

We should not underestimate the complexity of these disorders however. Claims by molecular biologists that within 20 years we will be able to identify genetically susceptible individuals and correct the defects are, even by the standards of *Daedalus*, extremely naïve. While some monogenic diseases may be controlled in this way, its application to polygenic disease is much more difficult to anticipate. Recent work on the non-obese diabetic mouse, a very good model for human type 1 diabetes, suggests that at least ten different genes may be involved, and possibly more. And several hundred genes are involved in lipid metabolism. It will take a long time to understand how the products of these different loci interact with each other, and with the ill-understood mechanisms of ageing, to generate our problem diseases.

Cancer is a good example of the difficulties which may face us. It is now clear that many cancers result from mutations of oncogenes, that is genes which are involved in regulating the way in which cells divide, differentiate, and mature. Colon cancer appears to require the acquisition of mutations in six or more oncogenes; a few cancers may require fewer mutations, but at least two. How all these mutations interact to result in a malignant phenotype is still not clear. And to what extent they reflect the actions of environmental carcinogens compared with endogenous DNA-damaging agents produced as part of the process of ageing also remains to be determined.

Thus I suspect that progress towards the control of our major killers and causes of chronic health will be slow. We shall gradually start to understand some of the genes involved and this may well direct us towards a better understanding of their pathogenesis. In some cases this may help us to devise better programmes for their management. And along the way we will undoubtedly acquire powerful diagnostic agents for determining susceptibility to these diseases or for their early diagnosis. Thus there will be a gradual improvement in their control and management but I doubt whether there will be a dramatic change in medical practice over the next 50 years consequent on this new information.

If the prediction that we will have reached a 'half-way house' in understanding and controlling our current killers is correct, what will medical practice look like in 50 years time? Preventative medicine will be a major force. Except for a few enthusiasts on the west coast of the USA it will not have managed to turn us into a race of hunter gatherers, though we will be leaner, more physically active, and our diets will be rich in foliage and equally unattractive nutrients. Cigarette smoking will have become an imprisonable offence, and pipes, so beloved of Haldane, will be museum pieces. Alcohol and addictive drugs will be difficult to obtain, but we will not have found the safer alternatives predicted by Haldane and Aldous Huxley. Regular screening of our blood, urine, stools, and anything else that is on offer will start at birth and continue to the grave. Some of our current killers and causes of chronic ill health will be preventable, or curable if they are recognized early in their course. But high technology medicine will still be with us, although it will look different to today's practice. Major surgery will be carried out through tiny incisions, monitored on television screens. Every organ, except the intact brain, will be transplantable, many of them coming from animal sources. Many acquired diseases, like cancer, degenerative disorders of the brain, and heart disease, will be treated by agents administered by gene transfer therapy, and drugs will be 'designer-made' from a detailed knowledge of receptors and their functions. While boredom, stemming from the spartan requirements of preventative medicine, will pose a transient psychiatric problem, it will gradually pass as generations are born which have never known the pleasures of cream teas, good claret, and a decent cigar.

Current epidemiological and demographic studies suggest that even if we are able completely to abolish the major causes of premature death in middle age, particularly heart disease and cancer, we will only add a few years to our life expectancies at birth. In effect, we seem to be programmed to survive for about 80 years. Is this likely to change? The 'disposable soma' theory of ageing suggests that we are adapted by our metabolic make-up to survive until we have reproduced, and what happens after that does not matter. We are, in effect, like the products of the motor car industry, designed to last in good order for a while, but not for too long. This notion is backed up by a considerable amount of biochemical evidence which suggests that we are continually producing endogenous chemicals which are capable of damaging our DNA and, as we age, we gradually lose the capacity to detoxify these agents and nullify their damaging effects. Although work on *Drosophila* and other organisms suggests that there are important genes involved in ageing, current progress in understanding these mechanisms suggests that we are unlikely to make a major inroad into controlling the rate at which we age over the next 50 years. On the other hand, this expanding field may provide valuable information about the role of ageing in generating some of our current killers, and for telling us how to avoid some of the more unpleasant accompaniments of ageing.

As we move towards the deadline set by the student's essay in *Daedalus*, we are likely to see a major change in our approach to tackling the problems of human biology and medicine. When we have a map or complete sequence of the human genome, and start to understand the actions of its individual genes, we may, in effect, be reaching the limits of how far we can advance by this type of reductionism. We will then be faced with the awesome problem of trying to find out how the whole thing works in an intact human being. At the moment we do not have the biomathematical skills to understand the coordinated activity of hundreds of genes in the smallest micro-organisms, let alone large multicellular organisms like man. But we will undoubtedly have to move in this more holistic direction if we are to make best use of what we are likely to learn about human gene action over the next 50 years, and, incidentally, to make genuine progress towards a real understanding of our current disease problems.

Perhaps the weakest part of *Daedalus* is its presumption that we shall learn enough about the complexities of human biology to be able to understand the physical basis for complex issues like behaviour, musical ability, bad temper, and intelligence. Despite some of the claims of the proponents of the Human Genome Project, I suspect that they, like Haldane, take a far too simplistic view of the factors which combine to produce the human phenotype. Although ideas reflecting this biological deterministic view of things are popular at the moment, they fail to recognize the importance of nurture in making us what we are. The kinds of people that we turn out to be must reflect the complex interaction of what we have inherited in our parental DNA together with the cultural environment which has been passed down from our forbears and in which we were reared by our parents. While I have no doubt that over the next 50 years we will learn a great deal about the way in which our genes determine what we are, it is highly unlikely that we will have reached the level of understanding of how these polygenic systems interact with an even more complex environment so that we can modify our personalities, talents, and political leanings.

MODERN BIOLOGY, MEDICINE, AND SOCIETY

Haldane was in no doubt that a great deal of what he had to say in *Daedalus* would be distasteful to many people. On the other hand, though he foresaw difficulties, he seems to have been convinced that our increasing ability to control our destinies would, overall, have a major beneficial effect on man's evolution and development. But was he correct in his apparent assumption that, in time, the applications of advances in biology and medicine would be accepted by society?

Almost every major advance in medical practice since Haldane's time has been followed by reactions ranging from concern, through despair, to furore. Modern high technology medicine has left us with a series of difficult ethical issues, spanning the definition of 'death' in a brain-dead person maintained on a life-support system to the establishment of priorities for health care in societies which cannot cope with its spiralling costs. But it is the artificial control of reproduction and genetic manipulation which, as evidenced by the plethora of books that have been published on ethical issues of biology and medicine over recent years, are causing most concern.

When the first babies were born after *in vitro* fertilization in the late 1970s there was great consternation. As this field developed, and research on the processes of fertilization and on the early embryo became more widespread, society reacted in a predictable way. In Great Britain attempts were made to provide an Act of Parliament to prevent embryo research of any kind. A committee was set up under the chairmanship of Mary Warnock to examine the whole question of *in vitro* fertilization and research on human embryos. The Warnock Committee took rather a pragmatic view of this sensitive subject which, while recognizing the enormous value of research in this field for the management of infertility, also accepted that to allow it to continue completely uncontrolled would be abhorrent to society. It made a number of recommendations, including the age up to which research on embryos could be carried out and practical issues regarding surrogacy and related questions, and concluded that *in vitro* fertilization should be regulated by a legislative body. This sensible compromise led to the establishment of the Human Fertilisation and Embryology Authority (HFEA). After a long period of debate in parliament this way forward was accepted, and for several years the field settled down and *in vitro* fertilization became widely available for the treatment of infertility. Hundreds of babies were born in this way, and at least for a while the topic ceased to be newsworthy. Other countries took a less liberal view, however, and in some cases research on embryos was banned.

At the beginning of 1994 several new developments were reported in the British press at about the same time which led to a national outcry and to a reopening of the debate on the whole question of the regulation of reproduction. It was announced that several women in their late 50s or early 60s had given birth to babies and, at the same time, that for social reasons a black woman had given birth to a white baby with ova donated from an individual from a different race. And the news broke that it was likely that, within a short time, it would be possible to obtain eggs from aborted fetuses or cadavers for helping infertile women to have children. The response to this flood of new information was immediate and highly emotional. The Secretary of State for Health appeared on television to say that 'women of 59 have not had babies over the centuries and in my view nor should they.' A 'shocked' French health minister announced that he was to legislate to ban 'medically assisted procreation techniques for postmenopausal women', and a leading gynaecologist said that international action should be taken to curb 'retirement pregnancies', and that fetal tissue grafting was likely to be ineffective. The HFEA hurriedly stitched together a 15-page consultation document which was released to Members of Parliament, church leaders, ethicists, doctors, and anybody else who was interested.

This type of reaction has followed many major advances in the biomedical sciences over the years. The first heart transplants were received in exactly the same way and yet, within a few years, this procedure became an accepted part of medical practice. It seems likely that the current advances in assisted reproduction will follow the same pattern. It is not surprising that society is shaken by the idea that a child in the future may have to be told that its mother was, in effect, the product of an aborted fetus. But this is little different in principle to a person having the heart of another who was killed in a road traffic accident. And while there may be many reasons why it is not advisable for couples to become parents for the first time after the age of 60 years, there may be individual cases in which the circumstances make this acceptable. The process whereby these subjects are widely debated and, where necessary, controlled by appropriate authorities, seems to have worked in the early days of assisted reproduction and is likely to continue to do so. None the less, it is essential that these questions are debated widely and that society sets its own norms, if necessary through its elected politicians.

Advances in human genetics, particularly following the advent of the new DNA technology, have also spawned a wide variety of new social and ethical issues. The prenatal detection of genetic disease and termination of pregnancy has been practised widely for many years; the DNA era has simply widened the scope for this practice. Although, overall, most societies have now accepted that parents should be allowed this option, there are still many who believe that prenatal diagnosis and selective abortion should not be permissible under any circumstances. And wider objections are sometimes raised. They centre round questions about whether it is appropriate for society to decide that physical disability is always a bad thing. Surely, it is argued, some of our greatest creative artists suffered from such afflictions. Do we want to terminate a pregnancy and lose a Beethoven? It does not seem very helpful to base our attitudes to the avoidance of genetic disease on such unusual individuals however.

There are numerous examples of outstandingly creative people who have remained in rude health for the whole of their lives. It is difficult to substantiate the argument that unusual talent, or even genius, is seen only in the light of serious disability. However, many thoughtful people still believe that society should care for its genetically abnormal children, however incapacitated they may be. On the other hand, equally caring physicians and others take the attitude that potential parents of genetically abnormal children should have the right to decide what kind of children they bring into the world, now that the technology is available to allow them to make this decision.

But while society has, by and large, accepted that it may be reasonable to terminate pregnancies for serious genetic diseases, there is still considerable concern about where the line should be drawn. Of course, what may be acceptable for one society may not be for another. For example, it is now current practice in China to terminate pregnancies for a relatively mild inherited blood disease. While this condition might not cause any serious disability for somebody leading a sedentary existence in an industrialized western country, it might be disadvantageous where the norm is only one child who is expected to spend its life in vigorous physical activity in the rice fields. But there is already a worrying tendency on the part of some clinical geneticists to advise that pregnancies are terminated for what appear to be relatively mild conditions. Where might this stop?

Genetic screening also raises a number of difficult ethical issues. Many of these problems were highlighted by the disastrous effects of the programme which was established for sickle cell anaemia in the United States in the early 1970s, which was not backed up by adequate education and a clear plan for what was to be done with the information which was obtained from the venture. Difficult ethical questions also arise when screening is carried out for conditions like Huntington's disease which only manifest themselves in later life. For example, to determine whether somebody is at risk of transmitting this disease it may be necessary to carry out a family study during which it may be discovered that certain family members will develop the disease later in life. Some may want to know because of the risk of passing it on to their children; others may find this information quite intolerable.

Our increasing ability to identify individuals at greater risk for heart disease, psychiatric disease, and other common killers or causes of disability in middle life also raises a number of worrying issues.

What will we do with this information during the long period in which we cannot use it to help our patients? Will potential employers or insurance companies demand to have information about our genetic profiles? What effects will this have on the cost of health insurance for example? As health care provision becomes increasingly commercialized, how can society protect those who, through no fault of their own, are found to have a genetic constitution which makes them more likely to succumb to disease in middle life? Currently we have no way of predicting with any degree of accuracy whether somebody will develop one of these common killers, but there are hints that this may not be the case for long. Recent work suggests that we may be able to anticipate the likelihood of breast cancer or the occurrence of Alzheimer's disease in certain subsets of the population long before we know how to use this information for the benefit of our patients. And what will we do if we think we have found genes which may make us more or less prone to develop schizophrenia or homosexuality, or other behavioural traits. How will society cope with this information? In particular, who will establish what should be accepted as 'abnormal'?

There is also concern about the practical issues arising from our increasing ability to manipulate and transfer human genes. In most countries, and after extensive debate, it has been agreed that somatic gene therapy is acceptable, provided it is adequately controlled. After all, it is, in effect, no different in principle to organ transplantation. The inserted gene will 'die' at the same time as the recipient and will not be passed on to future generations. At the same time, many societies have banned germline therapy, that is the insertion of genes into fertilized eggs which will then be passed on to future generations. It is felt that we are not yet in a position to alter the genetic make-up of our great grandchildren, who will have had no choice about the decision. Furthermore, now that it is possible to identify genetic disease in fertilized eggs it is unnecessary to develop germline therapy; the eggs can be sorted and only those that do not carry a defective gene replaced in a woman.

There is no doubt that the fear of a resurgence of the eugenic movement underlies many of the concerns about our increasing ability to control the mechanisms of reproduction and to manipulate our genes. At the time that Haldane wrote *Daedalus* he was an enthusiastic eugenecist; 'had it not been for ectogenesis there can be little doubt that civilisation would have collapsed within a measurable time owing to the greater fertility of the less desirable members of the population in almost all countries.' Thinking along these lines, first stimulated by Francis Galton's *Hereditary genius*, was widespread among British intellectuals in the early part of this century and the eugenics movement flourished, both in Great Britain and later in the United States and elsewhere. For example, by 1914 at least 30 states in the USA had enacted new marriage laws, many of which declared void marriages of idiots and the insane; others restricted marriage among the unfit of various types. The first state sterilization laws were passed as early as 1907, and over the next ten years similar laws were enacted by 15 or more states. These laws gave power to compel the sterilization of habitual or confirmed criminals or individuals guilty of offences like rape. Though some members of the British eugenics movement looked on with admiration, this type of activity never caught on in Britain to anything like the extent that it did in the USA.

As pointed out by Daniel Kevles in his excellent history of the subject, *In the name of eugenics*, serious geneticists like Haldane and Lionel Penrose became increasingly disillusioned with the mainstream of the eugenics movement. After the Second World War, and the news of the Nazi atrocities which had been carried out in the name of eugenics, it tended to peter out, but not entirely. For example, in 1969 Jensen published an article in the Harvard Education Review entitled 'How much can we boost IQ and scholastic achievement'. This paper posed questions about genetic variation in intelligence in different racial groups and suggested that the lack of performance among the American black population might reflect an innate lack of intelligence. And, as pointed out in a recent article by Evelyn Fox Keller, the survey of the life sciences published by the National Academy of Sciences in 1968, *Biology and the future of man*, contained statements about the betterment of mankind which were little different to those of the mainstream eugenecists of the 1920s. Indeed, Keller believes that, although the remarkable advances in molecular biology and human molecular genetics have taken as their theme the control of disease for the betterment of mankind, this is, in effect, simply a 'new eugenics' movement in which the term *normal* is becoming increasingly ambiguous, and in which the concept of individual choice is blurred. In some respects she is right to raise these issues. It is not reassuring to hear that in many Mediterranean countries, for example, testing for genetic blood disease is becoming mandatory before marriage. In societies in which cost-effectiveness seems to be the only yardstick for the provision of health care, there are increasing concerns that parents may be pressured into undergoing screening and termination of pregnancy for diseases which place a major burden on health services.

The recent announcement by the Chinese government that it is to introduce a new law designed to eliminate 'inferior births', by forbidding those with a hereditary disease from marrying for example, has caused great consternation in the west. This new policy is contained in a draft law on 'eugenics and health protection' which was presented on 20 December 1993 to the Eighth National People's Congress Standing Committee in Beijing by the Minister of Public Health. Under this law those with conditions such as hepatitis, venereal disease, or mental illness 'which can be passed on through birth' will be banned from marrying. There is a danger that if China's new policy extends to genetic screening programmes for serious monogenic diseases it will stimulate a backlash of opinion against the use of this technology in the west.

But if the western press had kept a closer eye on events in Southeast Asia over recent years it might not have been so surprised to hear about the recent announcement from Beijing. In 1988, China's Gansu Province passed laws designed to improve the quality of their population by banning marriages of mentally retarded people unless they had first submitted to sterilization. Similar laws have been adopted in other provinces. And in Singapore in 1984 Prime Minister Lee Kwan Yew, concerned about the relatively low birth rate among educated women, reasoned that, because their intelligence was higher than average, the quality of the island's gene pool might be suffering. For this reason he provided a variety of incentives to stimulate them to produce more children and, at the same time, to persuade the less educated to undergo sterilization after the birth of one or two children. The fact that a politician of Lee Kwan Yew's exceptional ability could be led into a fallacy of this kind is a prime case, if such is needed, for trying to ensure that our politicians of the future are scientifically literate, a problem which has been highlighted in the last few months in Great Britain by some of the statements from our political leaders on the question of the control of reproduction.

It appears, therefore, that eugenics, though its mainstream movement appeared to die with the announcement of the Nazi atrocities after the Second World War, is always in the background. And, as evidenced by the concerns expressed in Peter Medawar's 1959 BBC Reith Lectures, *The future of man*, even the most humane, logical, and liberal-minded among us are concerned with fears of a decline in the quality of the human gene pool. But when seen in its historical perspective the evils of eugenics have always reflected political extremism, social snobbishness, confused thinking, and, above all, flawed science. The medical applications of molecular biology, while they may have been overstated, will undoubtedly transform medical practice, and while they will raise many new ethical issues, to label them 'the new eugenics' is emotive and illogical, and does little to clarify the debate on where modern biology is taking us.

In effect, what we have been witnessing over recent years is that the speed of progress in the basic biological and medical sciences is too fast for society to take it all in. The problem is compounded by an appalling lack of scientific literacy, not just in the developing world but throughout the richer countries. In Britain, despite the fact that there are over 600 Members of Parliament, government was hard pressed to find a dozen or so to form a House of Commons Science Committee. The quality of debate on matters such as embryo research and the control of reproduction in the House of Commons was abysmal. Furthermore, our politicians continue to disregard the importance of teaching science in our schools. The result is that less children are applying to pursue science in universities, the quality of science journalism, with a few exceptions, is extremely poor, and we are breeding a society which is increasingly frightened of modern biology, in a large part owing to its complete ignorance of what is going on. Of course the recombinant DNA era has only been with us a short time and our current state of confusion will not last for ever. But it is vital that we open up an informed debate between scientists, politicians, and the public about where we wish to go as we gain increasing control over our futures.

Looking back over the years since *Daedalus* was written, it is difficult to identify many evils that have arisen from the application of genuine scientific advances, and the major travesties have resulted from the disastrous combination of flawed political thinking with flawed science. But this is not an argument for trying to set back progress in biological research; we must go on debating each step with the public and, if necessary, slow down from time to time until society is ready for what may appear to be an unnatural intrusion on human biology. For, as Peter Medawar has reminded us, if modern evolutionary theory has told us anything, it has made it abundantly clear that nature does not always know best.

POSTSCRIPT

In his preface to Ronald Clarke's biography of J.B.S. Haldane, Peter Medawar concluded as follows: 'people who are tired of reading how lofty thoughts can go with silly opinions, or of how a man may fight for freedom yet sometimes condone the work of its enemies, have a simple remedy: they need read no further. But they will miss a great deal if they don't.' This might have served equally well for an introduction to *Daedalus*. Even allowing for the age of its author, it is undoubtedly silly in parts, and shows little compassion for the concerns of its more sensitive readers. On the other hand, it reflects the extraordinary qualities of the man who wrote it. For as well as its brilliant insight into what the biological sciences would be capable of achieving, it provides lesser mortals with an exhilarating view of the possibilities for the betterment of mankind through genuine scientific advance. In just over half the time that Haldane predicted, the biological and medical sciences are entering the most exciting and productive phase of their long history. Undoubtedly there are problems ahead, but if there are a handful of scientists with half the ability of Haldane to popularize and explain their complexities to society, none of them should be unsurmountable.

SOME SOURCES AND SUGGESTIONS FOR FURTHER READING

Bock, G. and O'Connor, M. (eds.) (1986). *Human embryo research. Yes or no?* Tavistock Publications, London.

Clark, R. (1968). *J.B.S. The life and work of J.B.S. Haldane*. Hodder and Stoughton, London.

Haldane, J.B.S. (1924). *Daedalus, or science and the future*. E.P. Dutton, New York.

Haldane, J.B.S. (1985). *On being the right size and other essays* (ed. J. Maynard Smith). Oxford University Press.

Harvey, P. (1984). *The Oxford companion to classical literature*. Oxford University Press.

Jones, S. (1993). *The language of the genes*. Harper Collins, London.

Keller, E.F. (1992). Nature, nurture and the human genome project. In *The code of codes* (ed. D.J. Kevles and L. Hood), pp. 281–329. Harvard University Press, Cambridge, MA.

Kevles, D.J. (1985). *In the name of eugenics*. A. Knopf, New York.

Medawar, P. (1990). *The threat and the glory* (ed. D. Pyke). Oxford University Press.

Weatherall, D.J. (1991). *The new genetics and clinical practice* (3rd edn). Oxford University Press.

Wilkie, T. (1993). *Perilous knowledge. The human genome project and its implications*. Faber and Faber, London.

CHAPTER NINE

GENETICS AND THE FUTURE OF IMMUNITY

N.A. MITCHISON

Editor's note: The following chapter by N.A. Mitchison is different from the preceding chapters. Mitchison (nephew of J.B.S. Haldane) reviews the exciting prospects for future studies of immunity—a subject to which Haldane drew attention in *Daedalus*. Mitchison further speculates on the future of gene therapy as applied to immunity. Recent advances in these areas of biology would have interested Haldane greatly. Haldane was a pioneer in the development of the gene–enzyme and gene–antigen concepts in biology. This paper validates Haldane's prediction in *Daedalus* that biology would be the most exciting science of the future.

At first sight the future of immunity might seem to be a subject of limited interest. Immunity evolved simply as a means of defence against invasion by micro-organisms, a task which the immune system already performs pretty well. What more can one hope for? Why not leave our defence systems to fulfil their traditional role, aided where necessary by such well established practices as sanitation and vaccination. It would be a different matter if the human race were threatened by some new plague, but even the AIDS epidemic seems likely to respond eventually to traditional means of control.

This would be to take too narrow a view, for several reasons. First, the immune system does not do all that good a job. Tuberculosis never disappeared altogether even from the developed countries, and is now on the increase. Infections pose little threat to young and middle aged adults, but they do so still to infants and the old. For most of Africa, and for much of Asia and South America, the tropical diseases have never been brought under control. Infant mortality remains high, and the debilitating effect of chronic infection imposes a severe economic burden. Perhaps we expect too much of the immune system. In the course of ten years each of us may throw off 200 overt infections, in addition to the countless trespasses of micro-organisms across the surfaces of our gut and lungs. Yet it is the failure of immunity in the final and fatal infection that attracts attention.

Second, the risk of infection is increasing. The environment which the human race inhabits is changing faster than ever before. Increasing population density world-wide, the growth of megacities, rapid transport, and greater movement of peoples all provide fresh opportunities for our parasites. We know that as human beings we are entering a period of change without historical precedent. We cannot predict the consequences, but that is no excuse for not trying to think matters through and prepare ourselves as best we can.

Third, the actions of the immune system can themselves damage the body. Rheumatoid arthritis, multiple sclerosis, psoriasis, and perhaps a proportion of the dementias of old age seem to result from activities of this sort. Together these types of disease afflict as much as one tenth of the population of developed countries, and are responsible for a larger proportion of people in need of hospital care. These are the diseases which are grouped together in the category of autoimmunities, although for none of them is the underlying mechanism entirely clear. For none of them at present is there any real cure, in the sense that the disease can be halted or reversed, although sometimes the patient can be made more comfortable. Quite probably cures will be found which are at least partially effective, without the underlying mechanism ever having been discovered.

Neighbouring the autoimmunities are the various forms of immune damage generated by infection. For many of the tropical diseases it is these hypersensitivities which cause trouble, rather than the damage done directly by the parasites. Most peasant farmers in Indonesia are infected with small nematode worms, without ill effects. Occasionally the infection develops into elephantiasis, an enormous overgrowth of some part of the body in which the lymphatic vessels have become blocked as a result of a misplaced immune response. Overgrowth of the spleen or inflammation of the kidneys can develop in the same sort of way in malaria. Lyme disease is a recent discovery in the USA and Europe, which is caused by infection with a spirochaete. Most of the damage, particularly the long-lasting effects, results from misplaced immune response. The autoimmunities and these misplaced responses to infection blend into one another. One guesses that common infections play a part in triggering autoimmunity, although it has proved remarkably difficult to obtain direct evidence of their doing so.

Fourth, there are prospects of manipulating the immune system so as to make it perform new tasks, tasks which in its evolution it has never encountered before. Using the immune system to attack cancer is one example, and using it for purposes of birth control is another. It will not be easy to make the body carry out work to which it is unaccustomed. For these new tasks we shall need to create immunological illusions. The immune system will need to imagine that it is just doing another of its old jobs of defence, when in fact it is doing something radically new.

A fifth and last reason for interest in the future of immunity is for the insight it will provide into the working of complex physiological systems. Haldane predicted in *Daedalus* that the early twentieth century decades of chemistry and physics would give way to an era dominated by biology. Following the discovery of the structure of DNA in 1950 that is indeed what happened. Now that the framework of molecular genetics is firmly in place, I predict that the next decades will be dominated by a growing understanding of integrative systems. The working of embryonic development, of the nervous system, and of the immune system will come to be understood. Rules which govern these systems will emerge, and as they do so we shall become able to predict, manipulate, and cure. Some of the rules will be unique for each system, but some will be common, so that understanding one of these systems will help us to understand the others.

THE FUTURE OF VACCINATION

Vaccines within the limited range currently available are highly effective. The eradication of smallpox was a triumph of vaccination, and both measles and poliomyelitis are nearing eradication in the USA by the same method. The proportion of children receiving effective vaccines is slowly but steadily rising in developing countries. What remains is partly the problem of reaching children in need in very poor countries, and partly the lack of vaccines effective against the major tropical diseases. No vaccines are yet available against malaria or most of the other protozoan parasites, or against the worms which infest humans. In addition, it is likely that ageing people in developed countries could benefit from specialized vaccines against bronchitis and the other infections to which they are prone.

There are many promising new ideas for vaccine development. Malaria is the greatest killer of them all, and there it may be possible to vaccinate against the sexual form of the protozoan which causes the disease, which matures only within a mosquito which has sucked infected blood. The vaccinated individual would make antibodies which would pass into the mosquito's gut along with the maturing parasites, and would there destroy them. Vaccination would not directly benefit the individual, but he would contribute to eradication of the parasite: an altruistic vaccine.

More generally, it is hoped that vaccination with a single component of a parasite may be effective. Single peptides might be better still, if the different peptides of one protein compete with one another for the attention of the immune system as strongly as Eli Sercarz in Los Angeles believes. Such vaccines would depart far from the classical form of killed or attenuated whole organisms. They would be made by the methods of molecular genetics, just as the first vaccine of this type, against Hepatitis B virus, is already being manufactured.

The problem in vaccine development is not so much in identifying the parasite and its components, as in knowing how to measure the response. No trouble, you might think, in measuring protection against disease. Alas, testing in a tropical country is often prohibitively slow, costly, and inaccurate. The answer is to understand the changes in the immune system which we hope to bring about by vaccination, and then measure them directly, although that is easier said than done. The key to this problem, I believe, is to understand the nature of the immunological memory trace which vaccination establishes in the immune system. An important part of this problem has recently been solved, and it is worth going into detail about the solution because it so closely relates to the problem of autoimmunity which is discussed below.

The immune system contains two main types of cell, B cells and T cells. The main function of B cells is to make antibodies. T cells have more diverse functions: some act as killer cells which destroy other cells harbouring viruses, while others act as regulators producing cytokines which transmit messages to other cells (the targets of these cytokines may themselves be T or B cells, or other cells of the body). All this has been known for some time. What has only recently been discovered is that these two types of cell work quite differently from one another in an unexpected way, or at least in a way which was at first unexpected but becomes obvious once one thinks about it. The differences can be summarized as follows.

1. The genes which encode the antibody combining site in B cells hypermutate. On average they undergo approximately one mutation at each cell division, which is a rate several thousand-fold higher than that of T cells or any other known cell.

2. Hypermutation occurs in a compartment of the immune system, called a germinal centre, where an antigen (say a virus) is stored. B cells compete vigorously with one another for this antigen, for success in grabbing a part of it stimulates the cell to divide and multiply. Having an antibody better able to combine with the antigen enables a B cell to succeed in the competition. Thus memory in a pool of B cells is established by the selection of cells endowed with better-fitting combining sites, which were generated by hypermutation. For this process to operate successfully, it is obvious that each individual B cell has to be selectable. Entry into the pool of memory cells will be by *proportional representation*, as in a French or German election.

3. The value of this process is that during the course of time after immunization antibodies with very close-fitting combining sites come to be generated. Antibodies selected in this way can work at minute concentrations, as low as 10^{-12} molar. As antibodies in total are present in blood at a concentration of 10^{-4} molar, less than one molecule of antibody in a million can still work effectively. Accordingly, after selection has occurred the immune system can afford to reduce the number of B cells making the particular antibody to only a few, while still retaining a memory effective against the virus.

4. Unlike B cells, T cells do not secrete their antigen-binding receptors. Large numbers of these receptors therefore work in concert on each cell surface, enabling cooperative binding to take place so effectively that tight-fitting combining sites are not needed. Recent measurements find values of binding for T cell receptors at a level several thousand-fold lower than for antibodies. Accordingly, T cells do not hypermutate.

5. Memory in T cells has two components, one of transient cell activation, and the other of longer term expansion of the activated cells. When T cells are stimulated by an antigen, such as a virus as just mentioned, they divide and multiply in the same way as B cells, although probably to a lesser extent. In addition, they undergo a

change which has no counterpart among B cells. They acquire a set of new surface proteins, and at the same time change their internal signalling and response machinery in a way which is poorly understood. Consequent on these changes they become hyper-responsive, in that they respond more easily to a second stimulation with antigen. Eventually the cells lose these changes and return slowly to quiescence, a process which in man takes several weeks. During their period of activation the body is protected by the hyper-reactivity of its T cells, and later these cells provide further protection simply by being present in an expanded number.

6. Having less need to select individual cells, T cells do not operate according to proportional representation, but rather by what amounts to a *first-past-the-post* selection system, as in a British or Indian election. The immune system seems to be able to harbour small numbers of potentially reactive T cells which fail to participate in a response. They may do so simply because they are such a minor population, or perhaps they may be constrained by special mechanisms. The latter possibility has a powerful advocate in Irun Cohen of Rehovoth, and we return to it below. The absence of hypermutation, plus a selection system of this sort, clearly fits T cells well for the task of taking responsibility for tolerance of self. There is ample evidence that this is exactly what they do.

Having dissected immunological memory into these three strands, one for B cells and two for T cells, we can now return to the subject of vaccination. The contribution of each of these strands can be measured, and between them they should provide what is called a 'surrogate marker' of the efficiency of a vaccine. That is to say, instead of waiting perhaps indefinitely to find out whether a vaccine will protect, we should be able to assess the likelihood of its doing so quite rapidly. When developing a vaccine step by step, such surrogate markers are invaluable, and are already being used in one way or another. Although a full dissection as outlined above has never yet been implemented, it is surely part of the future of vaccination.

THE FUTURE OF AUTOIMMUNITY

One's hopes for autoimmunity largely depend on what one considers to be the underlying mechanism. Two alternative ideas are current at present. One, which traces back to an idea propounded by Macfarlane Burnet in Melbourne some 40 years ago, is that autoimmunity occurs when a self-reactive T cell accidentally enters the mature cell pool without having been eliminated in the thymus, as would normally happen. The alternative idea assumes that it is normal for self-reactive T cells to occur among mature cells, where they are held in check by some sort of regulatory constraint. The former could be termed the theory of developmental error, the latter the theory of perturbed regulation. Developmental error is the more pessimistic, for it is hard to imagine how a very rare event could be anticipated or prevented. It is sometimes advanced by immunologists who study the thymus as a justification for their research, which is a quite unnecessary piece of propaganda as there are so many other good reasons for studying this important organ.

Fortunately the evidence strongly favours the perturbed regulation theory. Let me cite two recent sets of experiments. In one, a number of different genes encoding supposedly opposing cytokines were knocked out by homologous recombination, mainly in the laboratory of Professor Rajewsky in Cologne. Whichever cytokine was knocked out, the mice eventually developed autoimmunity in intestinal tissue. In another, my student Susanne Schneider examined mouse T cells reactive with the self-liver protein 4-hydroxyphenylpyruvate dioxygenase. This enzyme is familiar to immunologists under the name of F liver protein. It is well known that mice do not respond to this self-protein even when immunized with it in the presence of powerful adjuvants, a fact which Schneider could confirm. This is probably because the protein leaks out of liver cells into the blood, where it is present at a low concentration (10^{-9} molar), and from there reaches the thymus. Schneider reduced this concentration still further, by repeatedly injecting mice with antibody to the protein (the antibody had been made by immunizing other mice with a genetically altered form of the protein, to which they were not tolerant). The mice treated in this way became able to make a detectable T cell response when immunized with the protein. She estimated the frequency of reactive T cells by hybridoma analysis, a procedure which we need not go into here in detail, and also estimated the goodness of fit of their antigen-receptors. To her surprise she found that self-reactive T cells were definitely present in normal mice, although their frequency was approximately five-fold lower than in the antibody-treated animals. Even more to her surprise, the goodness-of-fit of the receptors did not rise after antibody treatment. On second thoughts, however, these findings seem less unexpected, and indeed they are just what the first-past-the-poast theory outlined above would predict. In fact they constitute perhaps the best evidence to date of the ability of the immune system to constrain a minority of self-reactive T cells.

What could the mechanism of constraint be? This is an important question, for the answer might tell us how to cure autoimmune disease. When mending a car, it helps to know how the engine works. Evidently T cells make committee decisions, and they need to know how their fellow T cells are voting. One way that they could learn this would be via the receptor–anti-receptor network, a theory proposed by Nils Jerne of Basel. As well as recognizing foreign antigens, T cells can recognize and react to one another's receptors. One might suppose that this mechanism would lead to self-destruction of the immune system, but in practice it seems to operate only within certain limits, so that T cells end up exercising control over one another to only a limited extent. According to this theory, autoimmunity represents an error in this control mechanism, which should be treated by therapy aimed at re-establishing a stable network. Vaccination with T-cell receptors would seem to offer a reasonable approach to doing so. This stratagem has been tested in various animal models of autoimmunity with encouraging results, but so far it has not been possible to demonstrate a network of this sort operating more generally in the immune system.

I think it more likely that these committee decisions are taken through cytokines and their attendant adhesion molecules, as mentioned above. Cytokines are proteins much like hormones, except that they transmit information between neighbouring cells rather than through the blood stream. Within the body most T cells are clustered around dendritic cells, to which they adhere by specialized adhesion proteins. In each cluster T cells seem to communicate with one another by secreting cytokines, some of which encourage the neighbouring cells to divide and differentiate, while others act as inhibitors and do the reverse. There are many reasons for accepting the importance of this kind of communication. For instance, immunologists have long noted that 'suppression' can occur within the immune system but have been unable to agree about the mechanism involved. More and more of these manifestations of suppression are now being explained in terms of inhibitory cytokines.

As regards perturbed immunoregulation in autoimmunity, I am much impressed by a recent discovery of my colleagues Katharina Simon, Eva Seipelt, and Joachim Sieper. They found that T cells in the joints of rheumatoid arthritis patients are biased against producing the cytokine interleukin-4, while T cells in the same location in the sister disease reactive arthritis have the opposite bias. Systematic biases of this sort are revealing, although they do not tell us whether the distorted pattern of cytokine production is what causes the disease, or vice versa. That question can be answered only by therapeutic intervention, an enterprise in which we and others are engaged. But before looking at the future of intervention, let us consider a neighbouring area where the same kind of possibilities are much discussed, the immunological treatment of cancer.

THE FUTURE OF IMMUNITY TO CANCER

When considering the future of immunity one needs to bear in mind the wild fluctuations of optimism and pessimism of the past. During the 1950s the demonstration of rejection antigens (antigens strong enough to cause rejection of cancer transplants) in mouse cancer gave rise to high hopes, only to be dashed when the same antigens failed to materialize in human cancers. In the 1980s there was a revival of interest in BCG and other bacteria as immunological adjuvants for the treatment of cancer. Although highly effective in mouse cancer, they proved almost entirely ineffective in human trials, with the solitary exception of bladder cancer (possibly because this type of cancer harbours viral antigens). At the present time optimism is again reviving, thanks to a better understanding of the antigens against which anti-cancer immunity might be targeted.

By now we are well on the way to understanding why the cancer antigens of mice are so much stronger than those of man. My view is as follows. Cancer is caused by mutation of oncogenes, the genes which control the growth of cells, and which when mutated allow uncontrolled growth. Mutations occur at a low rate in all cells (other than the hypermutation of B cells mentioned above). When a sufficient number of mutations accumulates in oncogenes, a cancer develops. At the same time mutations occur at the same average rate in other genes, which vastly outnumber the oncogenes. Although these non-oncogenes play a lesser part in growth control, mutation in them tends to put the cell at a growth disadvantage, so that most of these mutations are screened out before the cancer develops. At least that is what seems to occur in most human cancers, which develop over a long latent period. Mouse cancers, particularly those induced by the powerful carcinogens which are often used for experimental purposes, have less time for these non-oncogene mutations to be screened out, so that more of the mutations are retained in the cancer cells. It is these mutated non-oncogenes which I believe encode the rejection antigens of mouse cancers. The group of Thiery Boon in Brussels has made an observation which strongly supports this theory. They cloned three genes which encoded mouse cancer rejection antigens, and found them to be distributed at random through the mouse genome.

Up to a point this is bad news for our hopes of immunity to cancer in man. Like all clarifications, however, it tells us where and how to start looking for the exceptions. Part of the future will be a hunt for human cancers which have not succeeded in screening out their antigen-encoding random mutations. Another important possibility is that these mutations may not be as random as we think. Indeed Robert Souhami at University College London has discovered that mutations in cancer cells tend to occur at hot spots in the genome. One way or another, I doubt whether we have heard the last of these random antigens.

Random antigens are not the only possible target for cancer immunity. Cancers grow from many different tissues, and their cells express differentiation proteins characteristic of the tissue from which they originated. They also tend to express in relatively large amounts those proteins which a rapidly growing cell needs, and which are expressed in much smaller amounts by their normal counterparts. All of these differentiation and growth proteins offer a potential target for immunological attack.

Thiery Boon's group has made another promising discovery. Many melanomas express the enzyme tyrosinase, which is characteristic of the melanocytes from which this type of cancer is derived. The group was able to isolate killer cells from a patient which recognized peptides derived from the enzyme. Their success in doing so may reflect the secluded location of melanocytes. Many years ago Rupert Billingham and Peter Medawar in Birmingham spread melanocytes taken from one guinea pig over a raw area on the skin of another guinea pig. They discovered that the melanocytes could heal in and spread out from the original area, growing within their host's skin. In spite of the fact that these cells were foreign they were often well tolerated, although at any time they could be bleached out by transplanting a piece of skin from the original donor. Melanocytes evidently dwell in an immunologically privileged site, as it is called. Hence there may be no normal self-tolerance of melanocyte proteins, and that may explain why killer cells recognizing tyrosinase can develop.

In a sense immunity to these differentiation and growth proteins would constitute a beneficial form of autoimmunity. What we try to achieve in one context is exactly what we try to prevent in the other. Indeed features of the immune response which occur detrimentally in autoimmunity will eventually be put to good use in the treatment of cancer. Take another of Sercarz's discoveries, antigenic spreading. In autoimmunity in experimental animals the immune response can be seen to start up against a single part of a single self-protein. Then in the course of time it spreads to other parts of the protein, and to other self-proteins, none of which would ever normally be recognized by the immune system. These normally non-recognized components of self are referred to as 'cryptic'. Cross-talk within the T-cell clusters mentioned above no doubt plays a part in this spreading to cryptic components. Sercarz imagines that in the future patients will be vaccinated against cryptic components of their own cancer cells, to which they would not be self-tolerant. This would be expected to elicit production of killer cells able to attack the cancer.

There is the obvious snag that this kind of autoimmunity could also attack the patient's normal cells. One can reasonably hope, however, that these would survive because they would express low levels of the auto-antigen. Furthermore, a telling fact that emerges from the birth control trials described below is that attack on an auto-antigen need not do as much damage as one might fear.

THE FUTURE OF IMMUNO-CONTRACEPTION

There is little doubt that the immune system never normally attacks cancer cells. If it did so, one would expect to find a higher incidence of cancer among immuno-compromised individuals. For instance people with congenital defects of the immune system, or receiving immunosuppressive drugs after kidney transplantation, should get more of the common cancers. They do not do so, apart from certain viral cancers which can be regarded as the exceptions which prove the rule. So in persuading the immune system to make such an attack, we are trying to make it fulfil a purpose for which it has not evolved. There is nothing inherently impossible about that, although we would certainly feel more comfortable with nature on our side.

Immuno-contraception is another task of the same sort. The immune system of man has evolved so as to avoid attack on our gametes, our embryos, and the hormones that support placentation. It has never been entirely clear how it manages to do so, as none of these structures are present in the body prior to reproduction, and all of them are therefore potentially targets for attack. Nevertheless it succeeds admirably in avoiding attack, give or take the rare case of immunological infertility. Immunologists have now set themselves the daunting task of developing vaccines against fertility.

The reasons for doing so are obvious. The world demographic crisis is so urgent that no possible technique of contraception should be neglected. An anti-fertility vaccine has much to be said for it. It promises to be acceptable, long-lasting, reversible (in the sense that its effects would wear off in the course of time without impairing subsequent fertility), free of side effects, and cheap. Pran Talwar's group in Delhi has gone a long way towards developing such a vaccine, which is based on the pregnancy hormone chorionic gonadotrophin. This vaccine has sailed through a series of trials, where it has proved safe and effective. The World Health Organization's Human Reproduction Programme has sponsored development in the USA of a somewhat similar vaccine, which has been less extensively tested. The next step is to solve the problems of large-scale manufacture of the most effective vaccine.

A curious but not so far alarming feature of the Talwar vaccine is that it elicits antibodies cross-reactive with another protein hormone, luteinizing hormone. This hormone is essential for menstruation, and blocking its action would put a woman into menopause. Talwar observed that women making these cross-reacting antibodies continued to menstruate normally. Evidently these auto-antibodies were not able to block function as effectively as had been feared. For our cancer theme, the moral is not to exaggerate the danger of autoimmunity from cancer vaccines.

None of the birth control vaccines so far tested has proved 100 per cent effective, as perhaps was to be expected. We need to think very carefully about the genetic consequences of widespread use of such a vaccine. Common sense tells one that a vaccine which is either very effective or very ineffective would have a lower genetic impact than one which is intermediate. A vaccine of intermediate effect could have enormous impact on the genes which control immunity in the human population. The delicate balance of HLA polymorphism would be particularly vulnerable to this form of attack, which would drastically select against those HLA types best able to respond to the vaccine. This does not of course mean that immunity as a whole would diminish, because the HLA system is highly selective in which responses it will support. The fact is that the genetic consequences could be enormous, and are quite unpredictable. In comparison, the population effects of genetic engineering, as discussed in the next section, would be trivial.

THE FUTURE OF GENE THERAPY AS APPLIED TO IMMUNITY

To a limited extent gene therapy is already being applied to the immune system. Cytokine genes are being implanted into cancer cells in the hope of enhancing the immune response to their antigens. So far this form of treatment has not yielded much clinical improvement, and the evidence of enhanced immunity has been at best equivocal. A more straightforward use of gene therapy is in replacing gene functions lost in rare congenital defects of the immune system. Successful treatment of adenosine deaminase deficiency is a case in point. Defects of this sort are well suited to this form of therapy because the target cells are stem cells. Both T and B cells are derived from stem cells in bone marrow, and to implant genes into these cells, whence they would spread into their mature descendants, is a reasonable aim. A further proposal is to help this spreading by implanting a selectable marker, such as a drug-resistance gene, along with the replacement gene.

All this is for the near future. In the more distant future lies the possibility of intervening in the main communication system, the cytokine network. Cytokine intervention could take various forms: use of cytokine-selective drugs such as thalidomide, perturbations which favour the production of particular cytokines, and the more direct procedures of administering blocking agents or recombinant cytokine proteins. Currently there is a flurry of interest in oral tolerance, a procedure which is believed to stimulate production of the cytokine transforming growth factor beta. For instance, we in Berlin and others elsewhere are feeding rheumatoid arthritis patients with collagen derived from cartilage; our trial is still blinded. Antibody to another cytokine, tumour necrosis factor, has yielded promising results in the treatment of rheumatoid arthritis in Marc Feldmann's trial in London. Implanting cytokine genes has not yet worked its way up the list, mainly because doctors are reluctant to use such a novel form of treatment before the other options have been exhausted. We shall probably have to pass through a preliminary phase of treatment with very expensive engineered proteins.

I am optimistic about the future of cytokine gene therapy in autoimmune disease for several reasons. First, the study mentioned above on rheumatoid and reactive arthritis suggests that these diseases may result from cytokine imbalance, brought about by lack of particular inhibitory cytokine proteins. This is precisely the kind of defect likely to be best treated by gene therapy. Second, provided that the gene can be implanted in the right place, the pharmacokinetics of gene therapy are likely to be favourable. That is to say that this form of therapy should be able to get the cytokine into the right place at the right concentration for the longest possible time. Third, gene therapy will be cheap, something which is useful in the preliminary animal experiments, and essential if the whole of humanity is to benefit. And fourth, the anatomy of the immune system is peculiarly well adapted to this form of therapy.

The last point requires expansion. The cells which secrete the important inhibitory cytokines are T cells, and the crucial place where they do so is in the cell clusters mentioned above. How then to get T cells implanted with cytokine genes into these clusters? Fortunately the best way of doing so is long established and very well known. It is the procedure of adoptive transfer, whereby T cells are taken out of the body, manipulated at will, and then returned. Implantation of cytokine genes would simply join the rank of manipulations which immunologists have long been performing, such as filtering T cells through glass wool or treating them with antibodies. All sorts of methods are available for controlling and monitoring the activity of these cells after adoptive transfer, such as treatment with further antibodies or with antigen. Until recently this procedure was used only in experimental animals, but now Philip Greenberg in

Seattle has begun to explore adoptive transfer in man for protection against viral infection.

What are the hazards of gene therapy? The regulatory agencies have dwelt at length on the hazards of viral activation and contamination. In my experience ordinary people are more interested in the danger of contamination of the human gene pool. It is worth remembering that the same question can be raised about any form of effective therapy of congenital disease, and one should remember also that every disease involves a certain level of genetic susceptibility. In any case the hazard is slight, except in certain instances for family members. Many of the congenital defects of the immune system are sex linked, where the individuals at risk from the rescued defective gene would be the grandsons of the treated male. Autosomal defects are less common, probably because most of the affected genes are lost through reduced fitness of carriers (eminent geneticists often get this wrong: they invoke ascertainment bias, forgetting that most immunologically crippled children are male). As Haldane argued, the autosomal defects which we do observe will tend to have beneficial heterozygous effects, in which case we need not worry too much about genetic contamination.

CONCLUSION

This article summarizes recent research in vaccine development, autoimmunity, cancer, birth control by immunological methods, and gene therapy. It predicts future developments in these areas, and attempts to assess the genetic hazards involved.

CHAPTER TEN

POSTSCRIPT

KRISHNA R. DRONAMRAJU

In spite of the pessimistic overtones of *Daedalus*, Haldane was a perennial optimist with regard to the benefits of scientific applications to solve the problems of humanity. Some of his writings of the 1920s and the 1930s, for instance *Possible worlds* (1927) and *The inequality of man and other essays* (1932), contain such unbridled zeal and enthusiasm that they seem to border on naïvety. Yet, it would be unfair to say that Haldane was unaware of the dangers of the applications of science and technology. Long before the suppression of mendelian genetics by Lysenko (with the support of Josef Stalin) in the Soviet Union, Haldane cautioned that the close relationship which existed in the Soviet State between the state and science might prove disastrous if scientific theories ran counter to the official doctrine (Haldane 1932). In *Daedalus*, we can discern a mild warning against official control of science when Haldane defined a 'eugenic official' as 'a compound . . . of the policeman, the priest and the procurer . . .'

Science and technology have become greatly complicated since 1923. The evils as well as the benefits of scientific applications are enormously magnified since Haldane's time. The pace of technology is so fast that the genetic betterment of the human species will be achieved by methods which Haldane could not have foreseen. It is a safe prediction to say that technological change (or progress, depending on one's point of view) over the next 70 years would be at least as different as it is in 1993 when compared with 1923, if not more. But what makes *Daedalus* a timeless commentary on the human condition is its emphasis on the great need to develop an adequate moral and ethical foundation to assimilate new technologies. Recent developments in genetic engineering and reproductive biology have only emphasized this need very greatly.

EDITOR'S PERSONAL NOTE

Daedalus opens with a personal statement by Haldane about his experiences in the first world war. I knew Haldane intimately during the years 1957–64, at first as his pupil in genetics and later as a colleague. Much of that period was spent in India but I travelled with him extensively to participate in scientific conferences in Europe and Israel. The years I spent with Haldane were his last. As his sister Naomi (Lady Mitchison) put it: 'He had come to the point where he was bound to try and see science as a whole, to calculate what might happen next, to develop a philosophy which would make sense of it. He had to look into the dark glass of the future' (see Mitchison 1985). My years with Haldane during that last chapter of his life were filled with great intellectual excitement and intense scientific activity. I met him when I was 20 and he was in his late 60s. In spite of the extreme naïvety of my youth and innocence, I knew at once that I was in the presence of a great mind. I was familiar with some of his papers and books in genetics but almost none of his extensive contributions to various other fields. And Haldane was kind enough not to overwhelm me with his great scientific work but allowed me to approach him gradually with my questions. Indeed he was modest to a fault when discussing his own scientific work with a young student. He preferred not to call us 'students' but his 'junior colleagues'. It was because of my own curiosity and because of my reading of various books and scientific journals that I slowly began to appreciate fully the breadth and impact of Haldane's intellectual contributions. I was aware of *Daedalus* during my first year with Haldane but did not read it until after his death. How I wish I could relive those years once again so that I could ask Haldane to clarify and answer my many questions or what he meant by a certain statement! For his part, Haldane was too busy directing my attention to the works of Charles Darwin, William Bateson, and other great biologists who preceded him or to some recently published work.

Starting thus towards the end of Haldane's life, I had been working my way backwards during the past 30 years, slowly and gradually discovering Haldane's numerous publications, correspondence, and other archives as well as personal reminiscences of others who knew him. It has been a unique experience. It is as if I started reading the closing chapter of a book and continue to catch up with the rest of the book over many years though not always in the correct chronological sequence.

HALDANE, CHRISTIANITY, AND ATHEISM

Early in his life Haldane's rebellion against authority included religion as well. His sister Naomi later wrote that they were brought up without religious beliefs. However, their mother took them to New College (Oxford) chapel regularly where they learnt enough scriptures to win prizes. She added: 'We had a set of strict ethical principles which were slightly harder to live up to because there was no supernatural sanction behind them' (Mitchison 1968). Later, Haldane commented that he 'developed a mild liking for the Anglican ritual and a complete immunity to religion'. While at Eton, Haldane attended Sunday services but later recalled that an over-enthusiastic proselytizing matron drove him into the ranks of the Rationalist Press Association. Indeed, Haldane's first detailed discussion of his theory of the origin of life was published later in 1929 in the *Rationalist Annual* to which he contributed outstanding popular articles for the rest of his life.

Although Haldane remained an atheist for much of his life, he took a great deal of interest in all major religions, often commenting on their creeds and practices in his popular essays. During his last years in India, he was attracted to Hinduism of which he wrote: 'At its highest level Hinduism is certainly more compatible with science than is any other religion' (Haldane 1961). A year later, in an article entitled 'Beyond agnosticism', Haldane (1962) concluded that 'as time goes on more and more people will accept atheism and man's mortality as working hypotheses. This will not necessarily mean the end of religion . . . But my own belief is that though the religions are all untrue they are concerned with something very important.'

In *Daedalus*, Haldane wondered whether a religion which could satisfy the scientific mind would ever arise. He believed that this could only happen if that religion openly admits the provisional nature of its mythology and morals. Several years later, in 1948, Haldane once again expressed similar sentiments in a paper entitled 'Atheism', when he addressed the Oxford Socratic Club. It was founded in 1941 with C.S. Lewis, the well-known author, lay Christian theologian, and English tutor (at Magdalen College), as its first president. The purpose of the club was to apply the Socratic principle, 'follow the argument wherever it led them', to one particular subject—the pros and cons of the Christian religion. J.D. Bernal was one of the illustrious scientists who addressed the club before Haldane. In his address, Haldane, after 'debunking' theism, proceeded to expound the dangers of theism. According to the minutes, 'Professor Haldane suggested first that atheism is more intellectually honest and less cramping than theism founded on insufficient evidence. . . . In general the effect of a religious morality is to hamstring man's moral development. The true path lay in the use of each difficulty as it arose as a stepping stone to the next stage of moral advance' (Como 1979, p. 165).*

True to his beliefs, when he was dying of cancer in 1964, Haldane turned for comfort not to any religion but to his scientific work. In his last essay for the *Rationalist Annual*, which was entitled 'on being finite', Haldane wrote: 'I should find the prospect of death annoying if I had not had a very full experience mainly stemming from my work . . . I doubt whether, given my psychological make-up, I should have found many greater thrills in a hundred lives. So when the angel with the darker drink at last shall find me by the river's brink, and offering his cup, invite my soul forth to my lips to quaff, I shall not shrink' (Haldane 1965).

As a young man I enjoyed much kindness and valuable instruction from Haldane. I continue to find him to be not only a man of great learning but also an exceptional and great human being.

REFERENCES

Clarke, A.C. (1968). Haldane and space. In *Haldane and modern biology* (ed. K.R. Dronamraju), pp. 243–8. Johns Hopkins University Press, Baltimore, MD. Reprinted in Clarke, A.C. (1972). *Report on planet three and other speculations*, p. 226. Harper & Row, New York.

* Arthur C. Clarke, the well-known popular writer of science and science-fiction, who first met Haldane in London in 1951 as the President of the British Interplanetary Society, has recently sent me the following comment on *Daedalus*: '*Daedalus* really is an astonishing book, and it's a humbling experience to see how one of the most brilliant minds of the century fared in his attempts to forecast the future. He made several remarkable bull's eyes . . . However, I agree with his conclusion: "There can be no truce between the science and religion."' Interested readers may consult Clarke's article, 'Haldane and space', for a summary of Haldane's ideas in space research (Clarke 1968).

Como, J.T. (ed.) (1979). *C.S. Lewis at the breakfast table and other reminiscences*, pp. 165–6. Macmillan, New York.
Haldane, J.B.S. (1927). *Possible worlds and other essays*. Chatto & Windus, London.
Haldane, J.B.S. (1932). *The inequality of man and other essays*. Chatto & Windus, London.
Haldane, J.B.S. (1961). The dark religions. *Rationalist Annual for 1961*. Reprinted in *Science and life, essays of a rationalist*, p. 145, Pemberton, London, 1968.
Haldane, J.B.S. (1962). Beyond agnosticism. *Rationalist Annual for 1962*. Reprinted in *Science and life, essays of a rationalist*, p. 157, Pemberton, London, 1968.
Haldane, J.B.S. (1965). On being finite. *Rationalist Annual for 1965*. Reprinted in *Science and life, essays of a rationalist*, p. 192, Pemberton, London, 1968.
Mitchison, N. (1968). Beginnings. In *Haldane and modern biology* (ed. K.R. Dronamraju), pp. 299–305. Johns Hopkins University Press, Baltimore, MD.
Mitchison, N. (1985). Foreword. In *Haldane: the life and work of J.B.S. Haldane with special reference to India* (ed. K.R. Dronamraju). Aberdeen University Press.

INDEX